The Copper-Clad World

by

Harl Vincent

The Copper-Clad World
by Harl Vincent

Copyright © 2023

All Rights reserved.

ISBN: 978-93-59322-90-2

Published by

DOUBLE 9 BOOKS

2/13-B, Ansari Road
Daryaganj, New Delhi – 110002
info@double9books.com
www.double9books.com
Tel. 011-40042856

ABOUT THE AUTHOR

Harl Vincent (October 19, 1893 – May 5, 1968) was the pen name of American mechanical engineer and science fiction novelist Harold Vincent Schoepflin. He was a regular contributor to science fiction pulp magazines. Vincent was born in 1893 in Buffalo, New York. He went to a technical high school before enrolling in Rensselaer Polytechnic Institute. Vincent dropped out of RPI before finishing his first year to marry. He married Ruth Hoff, with whom he had two children: a son and a daughter. Vincent worked at Westinghouse as a mechanical engineer, specialized in the installation and testing of complex electrical devices. Later, he worked as a sales engineer, eventually becoming the manager of a local steam division. Vincent began writing after discovering Hugo Gernsback's pioneering science fiction journal Amazing Stories. "The Golden Girl of Munan" was Vincent's first published story, and it appeared in the magazine's June 1928 issue. Vincent published around seventy science fiction stories during the next fourteen years. Despite the fact that the most of his work appeared in early science fiction journals, he was published twice in the general fiction pulp magazine Argosy. Despite the fact that he stopped publishing in the early 1940s, Vincent remained active in science fiction.

CONTENTS

CHAPTER I
INTO THE UNKNOWN

ADRIFT in space! Blaine Carson worked frantically at the controls, his jaw set in grim lines and his eyes narrowed to anxious slits as he peered into the diamond-studded ebon of the heavens. A million miles astern he knew the red disk of the planet Mars was receding rapidly into the blackness. And the RX8 was streaking into the outer void at a terrific pace—out of control.

Something had warned him when they left Earth; the Martian cargo of k-metal was of enormous value and a direct invitation to piracy. Of course there was the attempt at secrecy and the shippers had sent along those guards. His engineer, Tom Farley, was thoroughly reliable, too. But this failure of the control rocket-tubes, missing their destination as a result—there was something queer about it.

"Tommy," he called into the mike. "Find anything yet?"

"We-e-ll, something," the audio-phone drawled after a moment: "I'm coming up."

"What is it, Tom?" he asked when the engineer's round face appeared at the head of the engine room companionway.

Farley dropped his voice and his customary smile was gone. "I found the stern rocket-tube ignition jammed so it's firing continuously," he said; "and the others are all dead: won't fire at all. That's why she doesn't swing to the controls?"

"Can't you fix it? Lord, man, we're headed out into the belt of planetoids. We'll be wrecked."

"Nothing I can do, Blaine, without shutting down the atomic engines. Then we'd freeze to death and run out of oxygen. These ships ought to have a spare engine just to take care of the heating and air conditioning. I always said so."

"What happened to the ignition system?"

Tom Farley looked over his shoulder apprehensively. "Dirty work, Blaine," he whispered. "I'm sure of it. Tool marks on the breech of the stern tube. And there's one of those guards I don't like the looks of."

"Nonsense. The k-metal people know their men; they picked these three especially for the job."

"Who else could do it? There's only the five of us on board."

There might be something in what Tommy said, at that. A thing like this couldn't just happen by itself. And, come to think of it, one of those guards was a queer looking bird: dwarfed and hunch-backed, sort of, and with long dangling arms. It would be better to investigate.

"Get 'em up here, Tommy," Blaine said.

THE RX8 drove on and on through the uncharted wastes outside the orbit of Mars. None of the space ships of the inner planets ever ventured out this far, and Blaine knew there was grave danger of colliding with some of the small bodies with which the zone was infested. If one of those guards was the traitor he was risking his own neck as well as theirs.

Two of them entered the control room with Tom Farley, big, husky fellows of stolid countenance and armed with regulation flame-ray pistols and gas grenades.

"Where's the other, the dwarf?" Blaine asked, his suspicions mounting immediately.

"In his bunk," Tom replied with a meaning look. "He said he'd be up in a few minutes."

The pilot-commander addressed the guards. "Fellows," he said, "I suppose you know we're in a serious fix. The ship is out of control and we've missed Mars, where your metal was to be delivered. We're speeding out into the unknown, out past the limits of space-travel toward the orbits of Jupiter, Saturn, Uranus—God knows where. And my engineer thinks that one of your number has tampered with the machinery. Know anything about it?" Blaine eyed them keenly.

One of the guards, Mahoney, flushed hotly. "No, sir," he snapped. "At least Kelly and meself had nothin' to do with it. But we've been suspicionin' that little Antazzo ever since we came out. It's a peculiar way he has about him, the divil."

"You think he—"

AN incisive voice from the doorway way interrupted, "Never mind what he thinks, Carson. I'll do the thinking from now on."

At one man they turned to face the speaker. It was the guard, Antazzo, and he was clothed from neck to ankles in a garment of bright metallic stuff that shimmered with shifting colors like those of a soap bubble. A mask of similar stuff covered his face, and in each hand there was a weapon resembling a ray pistol but of strangely unfamiliar design.

Mahoney shot from the hip and his stabbing ray splashed full on the hunchback's chest—but harmlessly. That lustrous garment was an insulating armor; the traitorous guard should have been shriveled to a cinder at the contact. Antazzo laughed evilly as his own weapons loosed strange and terrible energies.

Tom Farley ducked, and Blaine watched in horrified amazement as the crackling streamers of blue radiance from the dwarf's pistols found their marks. Mahoney and Kelly, standing there, bathed for a brief instant in horrid blue fire: tottering, swaying, their mouths opened wide in a last agonized effort, to cry out. Tiny pinpoints of brilliant pyrotechnics flashing and exploding within the columns of blue fire. Then, nothing! Where the two husky guards had stood there was utter emptiness; not even a shred of clothing remained. The air in the control room became heavy and acrid.

"Antazzo!" White-faced and shaking, Blaine cried out in futile protest, "My God, man, what have you done? What does this mean?"

And then, in a blaze of rage, he was on his feet. Murder was in his heart as he set himself for a crashing charge that would sweep the beast from his feet. His own flame-pistol was missing; it was a case of killing this monster with his bare hands. Tom was circling, over there, cursing horribly. One of them would get him. Strangely, Antazzo had lowered the muzzles of his pistols.

ATERRIFIC punch, started from the floor, never reached its mark. Blaine saw a tiny puff of pinkish vapor that spurted from the bosom of that metallic garment. He was coughing and gasping; helpless. Muscles refused to do his bidding. With a moan he dropped into the pilot's seat, knowing that Antazzo's will compelled him. That gas had hypnotic powers. Mechanically, his fingers strayed to the controls.

And Tom—good old Tommy—he was under the influence of the stuff too, creeping there on hands and knees toward the engine room companionway.

Antazzo was talking. "We come now to the matter of instructions," he said. "You, Farley, will assist me in restoring the ignition system to normal. You, Carson, will keep to the controls and will lay a course to Jupiter as soon as the control rocket-tubes will respond. Understand?"

Tom growled reluctant assent from where he was crawling.

Strange, this hypnotic gas! Blaine's mind functioned clearly enough, yet he was utterly at the mercy of this madman's will—a robot of flesh and blood. "Jupiter!" he exclaimed. "Why man, it's nearly a half billion miles from the sun. Not habitable, either."

Antazzo had removed his mask and now smiled a superior smile. "We'll reach it," he said: "the RX8 is very fast. And it's not the planet itself we're bound for, but its second satellite. Io, your astronomers call this body, and it's a world sadly in need of this marvelous k-metal."

"But—but—"

"Enough!" The hunchback snarled his rebuke in Blaine's face and turned to Tom. "Come, Farley," he said, as if talking to a child, "we must get to work."

IN a daze of conflicting emotions, Blaine turned to gaze through the forward port when the two had left the control room. The RX8 was accelerating rapidly under the steady discharge of gases from the stern rocket-tube and had already reached the speed of one thousand miles a second. If one of those tiny asteroids, even one no larger than a marble, should meet up with them it would crash through the hull plates as if they were paper. His heart went cold at the thought.

Oddly enough, he found himself *wanting* to make this trip with the demoniac Antazzo. It was the effects of the pink gas. Even with the misshapen guard down there in the engine room the power of his will was effective. The devil must be an Ionian, he thought. But how in the name of the sky-lane imps had he reached Earth? How had he wormed his way into the confidence of the k-metal people? He must have been there several years, working to this very end.

There was a tinkling crash on the starboard side amidships; a screaming swish as something slithered along the side and caromed off into the void. One of those little planetoids. Probably no bigger than a pea, and luckily they had struck it glancingly. He wiped the sudden perspiration from his forehead.

Pressure on the directive rocket controls brought no response. Would they never finish with that ignition system?

A gleaming light-fleck segregated itself from the mass of stars ahead. At first he thought he imagined it, but a second examination, this time through the telescope, convinced him it was growing larger. Drawing nearer, it was, and resolving itself into a well defined orb that was directly in their path. Fifteen hundred miles a second, the indicator read now! They'd never know what happened when they struck.

"Tommy!" he bellowed into the mike. "Are you fellows ever going to finish down there?"

THERE was no reply for a moment, and the blue-white globe drove madly toward them. He consulted the chart. Pallas—an asteroid some three hundred miles in diameter. Not very big as celestial bodies go, but big enough!

"Just one minute now." It was Tommy's voice coming drearily, unnaturally through the audiophone. A minute! Ninety thousand miles! It seemed the asteroid was that close already.

Antazzo was in the control room then, and the effect of his mental dominance became more pronounced. Suddenly the dwarf let out a shriek of terror when he looked through the port and saw the brilliant body that now loomed so close. Blaine experienced a savage joy in the knowledge that the hunchback was mortally afraid.

"Latza! Latza!" In his fear Antazzo lapsed into his own tongue. Then, remembering, he shouted, "We're ready, Carson. Swing wide!"

The directive rockets answered to their controls now. Quick pressure on this, a swift pull on that, swinging the energy value to maximum, brought results. The little vessel groaned and shivered under the strain as a full blast from the forward tubes retarded them. Her hull plates twisted and screeched as the steering tubes belched full energy in swinging them from their course. They were thrown forward violently, though the deceleration compensators were working to the utmost.

Pallas swung around in their field of vision, and there was a fleeting glimpse of sun-lit spires of mountains, shadowed valleys, and mysterious crevasses from which clouds of steam and yellow vapor curled. Still it seemed they must crash against one of those slender pinnacles. Nearer it came like a flash; a dizzying blur, now, that drove directly in their straining faces.

And then, abruptly, it was gone. Already thousands of miles astern, the danger was past. Miraculously, they had escaped.

Antazzo laughed; a hollow mirthless cackle. His fingers shook crazily when he untwisted them from their grip on the port rail.

"Good work, my friend. Very good, indeed," he jabbered, his chin quivering in nervous reaction. "And now we carry on—on to Io."

Blaine Carson, almost wishing they had collided with the spire, set himself grimly to the task. He was powerless to refuse.

CHAPTER II
THE SECOND SATELLITE

WHEN, eventually, they swung into the orbit of Jupiter and headed in toward the enormous red-belted body, the two Earth men were heartily disgusted with the voyage and with themselves. Repeated doses of the pink gas—the ignominy of their utter subservience to the will of Antazzo—had worn them down no less than had the hard work and loss of sleep. Both were in vile humor. They endured the triumphant chatter of their captor in bitter silence.

"Over there, my friends," he said, pointing; "see? It is our destination. The golden crescent, Io, is something over a quarter million of your miles from the mother planet. See it? It is home, my friends; home to me and for yourselves in the future—if the Zara spares your lives. Lay your course to the body, Carson."

Blaine growled as he sighted through the telescope. Yet, in spite of his fury, he could not overcome the feeling of excitement that came to him when the powerful glass brought the satellite near. This body was like nothing else in the heavens. Antazzo had called it the golden crescent. Rather, it was of gleaming coppery hue, and now, as they swung around, it was fully illuminated—a brilliant sphere of unbroken contour. Smoothly globular, there was not one projection or indentation to indicate the existence of land or sea, mountain or valley, on its surface. It was like a ball of solid copper, scintillant there in the weak sunlight and the reflected light from its great mother planet.

Antazzo laughed over his absorption. "Looks peculiar to you, does it not?" he asked; "rather different from any of the bodies you have visited, you are thinking."

Blaine grunted wordless assent. The globe that was Io rushed in to meet them, growing ever larger in the field of the telescope. Now it appeared that there were tiny seams in the smooth surface, a regular criss-cross pattern of fine lines that looked like—Lord, yes, that was it! The body was constructed from an infinite number of copper plates, riveted or brazed together to form a perfect sphere.

"WHY, the thing's made of copper!" Blaine gasped. "Copper plates. It's a man-made world; artificial. But where are the inhabitants?"

Antazzo laughed uproariously. "Not man-made, my friend," he corrected, "but preserved by man for his own salvation. A dying world, it was, and the cleverest scientists in the universe saved it and themselves from certain death. What you see is merely a shell of copper, the covering they constructed to retain an atmosphere and make continuation of life possible—inside."

"Your people live *inside* that shell?" Blaine was incredulous.

"What else? We must have air to breathe and warmth for our bodies. How else could we have retained it?"

It was staggering, this revelation. The young pilot could not conceive of a completely enclosed world with inhabitants forever shut off from the light of the sun by day and from the beauties of the heavens by night. Yet here it was, drawing ever nearer, a colossal monument to the ingenuity and handiwork of a highly intelligent civilization who had labored probably for centuries to preserve their kind. A titanic task! Who could imagine a sphere of metal more than twenty-four hundred miles in diameter enclosing a world and its peoples? A copper-clad world!

They were coming in close now, and the gravitational pull of the body made itself felt. Blaine was busy with the controls, sending tremendous blasts from the forward rocket-tubes to retard their speed for a safe landing. The incredibly smooth copper surface was just beneath them, stretching miles away to the horizon in all directions.

The inductor compass was functioning. Evidently Io possessed as strong a magnetic field as did the inner planets. Antazzo now consulted a chart which he drew from his pocket, and examined minutely the surface over which they were speeding. Here and there curious designs were etched on the copper plates, and it was from these he determined their course. Obviously there was an entrance to this sealed-in world.

WHEN they had proceeded some two thousand miles in a northeasterly direction Antazzo gave the order to reduce speed. Off at the horizon there appeared a bulge in the copper surface, a round protuberance that resolved itself into a great dome-shaped structure as they drew nearer. A full two hundred feet it reared itself into the heavens, and Blaine saw a number of large circular hatches in its side that evidently covered air-locked entrances.

"You will land close by the dome, Carson," Antazzo barked, "and both of you will get into your moon-suits."

At his tone Blaine saw red. He realized on the instant that the effect of the pink gas had worn off and that he was his own master once more. All the pent-up emotions of the past few days were unleashed. If only he could get in one good punch. They might get away yet. There was plenty of k-metal to replenish the fuel supply. He whirled suddenly, muscles tensed. He faced the grinning hunchback—and was greeted by a breathtaking spurt of the pink gas. This time it was not merely a subjecting of his own will to that of the master but a complete hypnotism, a somnambulistic state. As in a dream he turned to the controls.

Now it came to him that the dwarf no longer spoke. He worked his will entirely without words; his evil mind possessed fully the mind of his victim. For Blaine Carson there was no further independent thinking. He was an automaton, a sleep-walker.

LIKE a detached and more or less disinterested observer, he saw that he had landed the ship. Then he noticed three dwarfs in bulky, helmeted moon-suits, shuffling clumsily across the copper plates. Hazily he knew he was with the others in an airlock; the hiss and the throbbing of pumps told him that. Under the great dome there was the latticework of a huge reflecting telescope; strange pigmy figures scuttled here and there, working at curious machines. There was the constant purr of many motors, the gentle pulsation of floor-plates beneath his feet.

With the moon-suit removed, he realized the atmosphere was fetid and stifling. A great pressure bore on his lungs, making breathing labored and difficult. And then they were in a lift that dropped into the depths of its shaft with dizzying speed. Antazzo's grin; Tom's eyes, dull and lifeless, floating there in the haze before his own—it was all a nightmare from which he must soon awaken.

There followed a period of complete unconsciousness of movement and of his surroundings. Light—light everywhere; a blue-white radiance that beat upon his unseeing eyes with unrelenting ferocity. Stabbing pains bored into his very brain, pains that carried with them an unspoken and unintelligible command. Why couldn't they let him alone; leave him to die in peace? He knew he was on his feet, swaying. There were voices, strident and guttural, and then by some magic the veil was lifted. His brain cleared and he saw that he stood before a dais where a much bejeweled and resplendently clad woman sat curled in the luxurious cushions of a golden seat. Chalk-white was her face and her lips crimson; amazing eyes, cat's eyes, pupils red-flecked and glittering, stared out at him.

"THE Zara," Antazzo whispered. "You will make obeisance."

Mechanically, Blaine dropped to his knees and touched his forehead to the floor. Tom Farley, over there, was doing the same, but Antazzo stood erect with arms crossed over his chest and head thrown back. The eyes of the Zara swept him contemptuously from head to foot. All was not well between them.

Blaine arose from his humiliating position at a sharp command from the hunchback. Tommy did likewise and the two exchanged sheepish looks. The effects of the pink gas were wearing off once more. They were in a large hall, obviously the throne room of a palace. Men-at-arms lined the walls on either side of the dais, and these were straight limbed giants with green-bronze skin and regular features—not at all like the deformed Ionian who had captured them and stolen the RX8.

The Zara talked rapidly in throaty gutturals, her fierce gaze directed at Antazzo and her brows drawn together in a scowl that could have but one meaning. She was displeased with the hunchback, displeased and furiously angry. What was it all about? Hadn't he brought home the bacon—the k-metal they were after? Blaine was nonplused.

Then Antazzo replied to the woman who was obviously his queen. His voice rose in shrill disagreement and his scowl was as fierce as the Zara's. Threatening her, he was, the nervy devil. He clenched his fists and raised his arms in an angry gesture, pacing the floor in his fury and thrusting out a pugnacious chin while he raved. This Zara woman rose higher in her cushions, and the look that flashed from those terrible eyes would have warned a less excited human, however justifiable his anger might be. But Antazzo was in too deep to draw back, that was plain to be seen. Blaine held his breath in anticipation of an explosion.

IT came then, that explosion, and in a way entirely unexpected and horrible to behold. The tiger woman uttered one fierce sibilant like the hiss of a serpent, a terrifying sound that silenced the hunchback and brought him stiffly to attention, mouth open and eyes bulging with horror. One of those unbelievably white arms stretched forth, threateningly tense, and a jeweled finger leveled itself at the rash Ionian. From it there flashed an intangible something that leaped to bridge the distance with the speed of light, something that screeched as it flew and crashed like breaking glass when it struck Antazzo's horrified face. In an instant he was on the floor, screaming and writhing in mortal agony.

The Zara watched with compressed lips and livid features as a host of black disk-like things covered the squirming body, spinning madly as if driven by atomic energy and emitting a myriad high-pitched tones like the angry buzzing of a swarm of bees. Antazzo's body shriveled as the things hummed on in their devilish work. Soon there was but a tiny heap of clothing with the angry black disks whirling and singing their song of hate. And then, in a puff of thick yellow vapor they were gone, their gruesome work completed. The odor of putrefaction lay heavy on the air.

Blaine shuddered and a fit of nausea twisted his vitals. It served the devil right, of course, but it was a horrible way to go. These damned Ionians, even to their queen, were bloodthirsty creatures. And what devilish ingenuity they had displayed in their development of weapons! His eyes were drawn irresistibly to the flaming orbs of the Zara.

She was actually smiling at him, this beautiful, heartless animal, not a smile of derision but one of deliberate allure. He felt the hot blood mount to his temples. A languid arm beckoned him to her side and the amazing creature settled back in her cushions with the drowsy, contented motions of a lazy feline.

"Watch your step!" Tommy hissed.

That warning was unnecessary. Blaine shook his head and backed away from the dais, an instinctive recoiling from a loathsome thing. The Zara saw and understood; and she went again into a black rage. She sat stiffly erect and called rapid orders to her men-at-arms.

The Earth men were surrounded instantly, their arms and legs pinioned by powerful hands, their feeble resistance overcome by the bronze giants as easily as if they had been children. Helpless and hopeless, they were borne from the room.

THIS was the end of the story, Blaine thought. Why this Zara woman had not made away with them at once was a mystery. Perhaps they were being reserved for an even more terrible fate than that of the hunchback. They were being carried along a dim-lit passage now, and Tom was cursing his captors in a never-ending stream of invective.

A metal door opened and then clanged shut behind them. They were dumped unceremoniously on metal tables that resembled those of a hospital operating-room on Earth. Woven bands, quickly adjusted by the bronze giants held them fast. Blaine turned his head and saw that Tommy was still struggling against the inevitable. A gag had been placed in his mouth; that was why he had ceased reviling the Zara's servitors.

The room was cluttered with elaborate and complicated mechanisms that Blaine could not recognize in the slightest detail, excepting that there were many banks of slender glass cylinders which bore some resemblance to the vacuum tubes used on the inner planets for radio communication and television. One of the bronze giants, he saw, was bringing a metal cap from which a cable extended to one of the strange machines. This cap was forced down over his head with a none too gentle pressure and he could see no more.

There came a sharp buzz from the machine and a million stinging needles drove into his brain. There was a moment of fleeting visions; strange places he viewed, and strange creatures parading in a fantasmagoria of delirium before his aching eyes. Voices, harsh and guttural, spoke in his drumming ears; voices that were dimly understandable, though uttered in the tongue spoken by Antazzo and the Zara. Then came sudden and merciful unconsciousness.

CHAPTER III
ILEN-DAR

WHEN Blaine Carson opened his eyes it was to stare at the blue-white radiance of an illuminated ceiling. He lay on a downy cot and it seemed he had just recovered from a long illness. Weak and sick, he turned his head listlessly to gaze at the ornate embossed designs on a wall of gleaming silvery metal. What place was this? His mind was wool-gathering; dim memories of unspeakable things struggled for mastery over a hazy consciousness. Suddenly then he remembered, and he sat up in his unfamiliar bed, senses acutely alert.

Across the room he saw a figure hunched in a chair; a twisted man-creature who was oddly like someone he had seen. Antazzo! But this one had none of the other's ferocity as he returned Blaine's stare. Rather, there was a look of deep concern in his ugly face. He came immediately to the bedside and looked at Blaine solicitously.

"I see you have recovered," he said. "It is good."

Blaine stared and stared. This creature had spoken in the language of the Zara's subjects, yet he understood his every word! It must be a dream, this impossible thing that had happened. And where was Tom? Abruptly he found that he was talking rapidly in this tongue of an alien race.

"Yes, I've recovered," he said, "and I'm amazed at what I find. How have I acquired this knowledge of your language? Where am I, and where is my friend? Can you enlighten me in these things?"

The other smiled. "I can, Earth man," he replied. "You have been taught our language while you slept. A thought transference process we use for educating the young. You are in the palace of the Zara and your friend is safe in the next room. I may add that you are in high favor with Her Majesty."

The wizened creature lowered his voice on the last words, and his knowing eyes spoke volumes. In favor with that she-devil! Blaine went cold at the thought.

"I want to see my friend," he said shortly.

"Later. My orders are to bring you to the Zara immediately you are strong enough. And Pegrani obeys orders."

NO use to attempt a break now. Blaine was tempted to drive a fist into that ugly countenance and fight his way out of the place. But that would be suicide. He'd wait, get the lay of the land first and then try to dope something out with Tommy.

"All right, Pegrani," he said, "I'm ready to go before this Zara of yours."

As he prepared for the audience, alien thoughts crowded one upon the other in his strangely enlightened mind. With the knowledge of the language had come knowledge of many things relative to the copper-clad world. They'd given him a liberal education. Somehow he knew these stunted creatures like Antazzo and Pegrani were known as Llotta and that, while ruling the sealed-in planet, their kind had originally come from Ganymede, the fifth satellite of Jupiter. Centuries had passed since the inhabitants of Europa and Ganymede had been forced to desert their aging worlds and had settled on Io. During other centuries the widely different peoples had co-operated in constructing the great copper enclosure in order to keep the new world alive and capable of supporting life. Then had come a century of bitter warfare in which the Llotta were victorious. Intense hatred existed between the two races, he knew, and a hazy impression of mechanically imparted knowledge told him that few of the Europans remained alive.

"We are here, Carson," his guide announced, when they stood before the square columns of an enormous portal.

The scene in the throne room was vastly different than when he had first visited it. The Zara sat curled as before, a golden bowl of incense burning at either side of the throne. The men-at-arms were absent and, instead, there were dozens of handmaidens, white-skinned and seductive as their queen, reclining on luxurious cushions that were arranged in a semicircle before the dais. It was a scene of Oriental splendor. A stage carefully set.

PEGRANI knelt and touched his forehead to the floor but Blaine held himself stiffly erect, looking straight into the eyes of the Zara. She smiled and extended her arm in that beckoning gesture.

"You may leave now, Pegrani," she said, without deigning him a glance. "Remain in the corridor until I send for you."

There was a tense silence as the Zara's gaze, ineffably softened now, held Blaine's. Unconsciously he was drawn to the steps of the dais. Unwillingly, yet inexorably, his lagging footsteps brought him to her side. Cool white fingers touched his arm and he saw that the red flecks in the black of those wide eyes were golden now. Surely there was no harm in this woman. But he remembered Antazzo.

"Carson," she purred, "you are more than welcome to Llotta-nar, the land of my people and the ruling power of Antrid, the body you call Io. The freedom of the realm is yours for as long a time as you wish to remain."

This was too good to be true. "You—you mean," he stammered, "that Antazzo exceeded his authority in his act of piracy—in bringing us here?"

The golden flecks flashed red and a cold note was manifest in the throaty voice. "Antazzo," she replied, "was destroyed for his audacious actions. We needed this k-metal of yours, Carson, and he was sent to Earth to get a quantity of the material. By magnetic directional waves was he sent—we have no space-ships—his body disintegrated by my scientists for transmittal, and the atoms of his beastly form reassembled in their proper relation when he arrived there. But he threatened me when he returned successful. The possession of the k-metal and his knowledge of its powers and uses had gone to his head. He demanded my hand in return for his work; demanded that he be permitted to mount the throne of Llotta-nar as my consort. Therefore I destroyed him." The hard eyes softened anew. "And—and for his abominable treatment of you I destroyed him," she concluded.

BLAINE fought off the spell of those gold-flecked eyes; he looked away in sudden panic. This creature was not telling the truth. She was hiding something; a sinister motive lay beneath her smooth speech.

"My friend," he said abruptly: "what of him?"

"For your sake, my Carson," she purred, "he too shall have the freedom of the realm for as long a time as is desired."

The cool fingers crept along his arm, firm and compelling. "Look at me," she whispered.

He thought of the pink gas as his eyes were drawn irresistibly to hers. What he saw in those gold-flecked depths sent a shiver of apprehension chasing down his spine. Savage, devastating desire mingled with ill-concealed rage at his coldness. This beautiful animal could turn like a flash and rend him limb from limb—and would on the slightest provocation.

A commotion in the corridor caused her to release him and sit bolt upright. Temporarily relieved, Blaine wheeled to face the portal. Tommy had broken loose! He heard his strident voice, berating an unseen antagonist in the tongue of the Llotta.

Then they were in the room, Tommy struggling and arguing vociferously with one of the green-bronze guards. The handmaidens had deserted their cushions and were milling about in affrighted confusion. The Zara's sibilant exclamation startled him into looking at her once more. The same cold fury that had greeted Antazzo glinted icy-hard in that grimly beautiful face. It was all over for poor Tommy.

BUT the Zara reached upward and stroked a transparent rod that dangled above the throne, something he had not noticed before. A screaming vibrant note smote the heavy air, a pulsation that beat at the ear drums with painful intensity. Silence fell as the awesome sound died away and echoed faintly from the huge columns that supported the arched ceiling. Tommy cooled off when he saw that Blaine was unharmed.

"Drekan!" The Zara's voice was a whiplash as she addressed the guard. "You will leave my presence and report to your overman for punishment. Never again molest the Earth men. Begone!"

Again this amazing woman curled in her cushions and again she purred. Tommy watched in open mouthed astonishment as she smiled guilelessly on his friend.

"You may leave me now, my Carson," she cooed. "Farley is free to accompany you. Pegrani will guide you and inform you regarding our customs and our people. You will learn much. And then you shall return to Zara Clyone."

Blaine had fully expected that Tommy would die a horrible death before his eyes, and in his sudden relief bent low and kissed the cold white hand of the Zara. A foolish thing to do! She purred and snuggled into the cushions like the feline she was—a dangerous animal; claws drawn in now but ready to strike out, razor sharp, on a moment's notice.

PEGRANI led them along the corridor to a lift. The car shot upward with breath-taking speed.

"Say!" Tommy was growling, in English. "What's the big idea? You've got the old girl ga-ga. Trying to vamp her into letting us off easy?"

"Shut up!" Blaine returned, irritated. "I don't know where we stand any more than you do. But we're going to sit tight now and see what happens. No more rough stuff from you, either."

"What! You're going to just stand around and take it—whatever they hand us?"

"Of course not. But the time isn't ripe yet. We'll have to wait till we know what it's all about."

They were outside then, on the palace roof, and Pegrani motioned them to a railed-in runway that circled its edge. High overhead was the shadowy blackness of the copper shell that enclosed the satellite. Huge latticed columns, line upon line of them, stretched off into the distance as far as the eye could follow; enormous white metal supports that carried the immense weight of the covering which retained the dense and humid atmosphere. Myriads of tiny blue-white suns there seemed to be, stretching off between the columns, carried on thick cables and radiating the artificial daylight of the interior. Hot, damp odors wafted across the roof, the odors of decayed vegetation.

Most amazing of all, were the dwellings. In orderly rows like the columns, they were flat topped cylindrical things that reminded Blaine of nothing so much as the tanks of an oil refinery back home. And the space between was overgrown with dense tropical vegetation, tangled and matted and shooting transparent tubular stems up to a height of a hundred feet or more where they sprouted great spherical growths that looked like enormous sponges. Of a sickly, pale green hue, these growths overran everything; climbed the columns and were lost in the shadows above the multitude of lights. The big sponge-like blossoms expanded and contracted rhythmically. Breathing, they were, like living things. Specially cultivated plant life to assist in maintaining the oxygen supply balance by decomposition of carbon dioxide. A marvelous artificial world!

"THE streets and moving ways are in tunnels beneath the soil," Pegrani was explaining. "What lies before you is the city of Ilen-dar, capital city of the empire, and like all other cities of Antrid, it is self-sustaining. The vegetation is inedible, all of our food is synthetic and highly concentrated. You were fed by intravenous injection while under the influence of the language machines. Our heat and power is obtained from the internal fires of Antrid, and, alas, these are being exhausted with great rapidity. Our shortage of power is becoming acute, and again our peoples are facing extinction."

That explained their need for the k-metal. It came to Blaine in a flash that Antrid was in sore straits and that this expedition to Earth had more back of it than had been revealed. Even with the supply of k-metal Antazzo had stolen, they could not carry on forever.

A screaming object went hurtling through the blackness over their heads. Something, a vehicle of enormous size with rows of lighted ports on the under side, that roared its way under the roof of copper and was gone in an instant.

"One of our monorail cars," Pegrani told them: "a complete system interconnects all cities and divisions. They are capable of circling the globe in a day of your time."

Their familiarity with conditions on Earth was astonishing. Probably Antazzo was but one of many spies who had been sent to the inner planets. Pegrani discussed the speed in their own terms.

SOMEONE had crept up behind them; a slight, olive-skinned youth who touched Blaine softly on the shoulder. Pegrani did not see. He was pointing into the distance and expounding on the merits of the monorail system. The youth touched a finger to his lips to enjoin silence, and thrust a crumpled ball of metal foil into Blaine's hand before the pilot realized his intention. A message, undoubtedly!

Some instinct, or some slight sound, warned Pegrani and he turned on his heel just as the slender lad was slinking away. Black rage contorted his features and Blaine saw him make a quick motion toward the inner folds of his jacket.

"Pegrani!" he shouted as he saw a glint of steel. "Don't!"

But it was too late and the Llott paid him no attention, anyway. One of those wicked ray pistols sent forth its crackling blue flame and the youth stood there, bathed in the eery blue light; dazzling blasts of exploding atoms were seen within the flare. Then there was the nothingness into which Wahoney and Kelly had gone.

Blaine shouted horrified and angry protest and Tommy rushed in to mix it with their guide. But the glowing ray pistol waved them back. Other guards—the big green-bronze ones—were running in their direction.

"The message!" Pegrani snapped. "Give it to me."

Quick as a flash Blaine crumpled the foil more tightly. A hard little pellet now, he tossed it over the rail far into the matted vegetation below. One might as well hunt for a needle in a haystack as for that tiny ball. But Pegrani would not forget; he'd report to the Zara. They were in for it now.

CHAPTER IV
BEFORE THE COUNCIL

PEGRANI lost no time in reporting the incident to the Zara. The Earth men were hustled to the throne room of the palace where the leopard woman sat in conference with her advisers. An ominous silence greeted their entrance. Ugly faces leered at them from the long table.

"What is it, Pegrani?" The Zara's chalky face went whiter still.

"The Rulans, Your Majesty. They have endeavored to communicate with the prisoners."

"Did they succeed?" Clyone's voice was terrible in its fury.

"They did not. I destroyed the messenger, and the message itself was lost in the jungle where Carson flung it."

The Zara shot a fleeting glance in Blaine's direction and permitted herself the ghost of a smile. "It is well," she breathed. "But it must not happen again. Have Tiedor brought to me."

Pegrani hurried off to do her bidding and Blaine turned uncertainly to follow.

"You will remain, Carson—you and Farley." The incisive voice of the leopard woman halted him in his tracks.

Tiedor was chief of the Rulans, it developed. There was but a handful of them in the realm and they were the last survivors of the civilization of Europa; descendants of those original brave souls who had settled on Io as a last resort in the effort to perpetuate their kind.

He was a magnificent creature, this Tiedor, tall and straight in his muscular leanness and with wide-set gray eyes in the face of a Greek god. Olive-skinned like the messenger, he was, and with the high forehead of an intellectual. He swept the assemblage with a haughty gaze when he faced the Zara.

"TIEDOR," she snarled, "it has come to my ears that a Rulan lad carried a message to one of my guests from Earth. What means this?"

"I know nothing about it, Your Majesty." Tiedor gazed into the wicked eyes, unafraid.

"You lie! There is some treasonable scheme in which you had hoped to enlist their help. You will tell me the entire story, here before the council."

"There is nothing to tell."

"You will confess or I shall destroy every Rulan in the Tritu Nogaru." The Zara's words were clipped short with deadly emphasis.

Tiedor paled and his lips tightened in a grim line, but he stood his ground. "I have nothing to confess," he said.

With a whistling indrawn breath, the leopard woman threw back her head and motioned to one of the green-bronze giants who guarded the entrance. There was a nervous stir around the council table.

At her command the guard drew back a heavy drape that hid an embrasure in the far wall. There, on a stubby pedestal, was revealed a gleaming sphere of crystal, a huge polished ball that shimmered a ghastly green against a background of jet.

Slowly in its depths a milky cloud took shape, swirling and pulsating like a living thing. Then it flashed into dazzling brilliance and the globe cleared to startling transparency. It was as if it did not exist. Rather they looked through an opening in the cosmos that carried their gaze to another and distant point. It was a large open space that was revealed to their eyes; a sort of public square where many of the olive-skinned Rulans were coming and going to and from the entrances of the circular tank-like structures that surrounded the area. They were greeting one another in solemn fashion as they passed and watching furtively the green-bronze guards who were everywhere. The sound of their low voiced conversations came clear and distinct from the depths of the crystal sphere.

"Your choice, Tiedor," the Zara hissed.

"There is nothing—nothing, I tell you!" The Rulan chief's voice was panicky now.

CLYONE'S snarling command was carried to those guards out there in the Tritu Nogaru by some magic of the crystal sphere. As one man they snapped to attention. With deadly accuracy they released the energy of their ray pistols. It was a shambles, that square of the Tritu Nogaru; a slaughter house. Agonized screams of the doomed Rulans rent the air of the council chamber. They organized hastily and rushed again and again into the crackling blue flame of the disintegrating blasts of the guards' fire. It was hopeless: unarmed and unprotected, they were at the mercy of Clyone's minions.

Sick and trembling, Blaine cried out against the massacre. He was seized instantly by two of the green-bronze guards who had been watching his every move. Tommy, too, was in their clutches once more, fighting valiantly but without avail. The sphere went blank and silent, and the drape was returned to its place. Still muttering disapproval, the members of the council gazed at their queen in alarm. There was no telling what this vile creature might do.

"The slaughter continues. Tiedor," she gloated. "Soon your handful of followers will be no more. And good riddance."

Swaying drunkenly, eyes glazed with the horror of the thing. Tiedor went raving mad. In one wild leap he was upon her, his fingers sinking into the white flesh of her throat. Woman or no woman, he'd have her life.

But it was not to be. A quick move of jeweled fingers was followed by a crashing report. Tiedor staggered and drew back, spinning on his heel to face them all with distended, pain-crazed eyes. Astonishment was there, and horror, but the fire of undying courage remained. His olive skin turned suddenly purple, then black from the poisoned dart that had exploded in his entrails. He collapsed in a still heap at the feet of the Zara.

She stood there a moment in the awful silence, caressing her bruised throat with fluttering fingers. She had faced death for one horrid instant and was obviously shaken.

Then she recovered and flew into a rage. "Out of my sight, all of you!" she screamed. "Out, I say! The Earth men are to be freed and Pegrani will conduct them to their quarters. Go now!"

The councillors made haste to comply, jostling one another in their anxiety to jam through the doorway. Blaine found himself released. He took one step toward Clyone, murderous hatred in his heart. But he recoiled from the expression in those red-flecked eyes; they softened instantly and looked into his very soul, saw through and beyond him into some far place where relief and happiness might be attained. And then, suddenly, they were swimming in tears. The Zara dropped into a seat and buried her sleek coiffured head in outstretched arms, her shoulders shaking with sobs.

An incomprehensible anomaly, this queen of the Llotta; a strange mixture of cruelty and tenderness, of cold hatred and the longing for love.

A dual personality hers, susceptible to the deepest emotion or to utter lack of feeling as the mood might dictate.

Blaine tiptoed softly from the room.

THEY were in the corridor now, and Tommy was blowing off at a great rate. Even Pegrani was stunned and shaken. But Tommy raved.

"Forget it!" Blaine growled. "Where do we go from here?" He couldn't have explained his emotions then, even to himself.

"To our quarters, she said—damn her!" Tom Farley swore in picturesque English. "And we," he wound up his expressive tirade, "are getting in deeper and deeper. We can't do a thing. Why in the devil doesn't she put us out of the way and get it over with? What's she keeping us around for, anyway?"

Blaine was asking himself that very question. Pegrani regarded them with something of understanding in his beady eyes. But he was nervous and apprehensive and broke in on their conversation to urge them into action. The Zara must be obeyed.

The corridor was deserted now and their footsteps echoed hollowly from the bare metal walls. Pegrani was ahead, leading the way, when Blaine was startled by an insistent tap on his shoulder. Another of the Rulans, it was, repeating the gesture of the youth who had been killed on the roof. But this one had no message; he was after something else—telling them in pantomime to make a break for freedom and to follow him.

Blaine caught Tommy's attention. And Pegrani, warned again by that sixth sense of his, turned his head. With a bellow of rage he whirled into action, ray pistol in hand. But Blaine was prepared for him this time. He wasn't going to witness another murder—not now. Flinging Tom Farley aside, he let loose a terrific jab that landed full on Pegrani's mouth. The ray pistol crackled harmlessly, its deadly energy spending itself in searing the metal of the ceiling.

THEN he wrenched the weapon from the astonished Llott and was boring in with body punches that quickly had the dwarf gasping for breath. These creatures knew nothing of fighting with their hands except in the fashion of clumsy wrestlers. The thud of hard fists against yielding flesh was a new and terrifying experience. Pegrani was game, though, and he flailed about with his powerful arms, endeavoring to get his opponent in his grasp. Sidestepping to avoid one of his rushes, Blaine brought up a terrible uppercut that ended flush upon the Llott's jaw. His head snapped back and his knees gave way beneath him. Down he went in a flabby heap. Suddenly ashamed, the young pilot turned to the Rulan.

Tom's eyes were shining. It was easy to see that he felt better about things now.

"I am a friend," the Rulan whispered in the Llott tongue, "sent by one who would have conversation with you. It is of the highest importance, but we must make haste. Will you trust me?"

Blaine saw deep concern and sincerity in the fellow's blue eyes. "What do you say, Tommy?" he asked, looking to his friend for approval.

"I say, let's go. He seems okay to me."

Their new guide was familiar with the passages and especially so with dark and little used stairways that connected the floors of the huge building. They soon reached the roof through a hatch that opened on a small penthouse which was in deep shadow and entirely hidden from the runways where the green-bronze guards paced constantly.

A slender cable dangled before them, and at its lower end they saw a basket-like car which their guide bid them enter. When they had done so, he tugged on the cable, giving a rapid twitching signal. Instantly they were soaring up into the blackness above the lights of Antrid.

THE swift journey ended in a tiny enclosed vehicle where another Rulan operated the cable drum which had made the trip possible. The car was unlighted save for the faint glow of a hand lamp, and it was not until the lower door was closed that they were permitted a view of the interior of the strange vehicle and had a good look at the two Rulans.

"Now," the one who had brought them said, "I can explain. I am Tiedus, son of Tiedor. My companion is Dantus, son of Dantor, the greatest scientist in all Antrid. We are taking you to Dantor who has knowledge of the mad plans of the Llotta and is in need of your help in thwarting them. You are willing?"

"Why—why, yes," Blaine stammered, looking deep into the earnest eyes of Tiedus. "You—you know of the fate of Tiedor?"

"I do." The young Rulan fell silent; then shook his head as if to clear it of unwelcome thoughts. "There are but few of us left, oh Earth man," he said then, "and all expect a like fate sooner or later. But that is beside the point. We have important work to do: work that brooks no delay. We leave now for the Tritu Anu, with your consent."

Tom Farley was examining the machinery of the car with interest. "This one of the monorail cars?" he inquired, when Dantus had seated himself at the controls.

"Indeed not. The Llotta do not even know of the existence of this vehicle. We could not get right of way on the rails, so this gravity car was developed in secrecy. It is provided with variable repulsion energies that can be adjusted to keep it at a fixed distance from the inner surface of the copper shell. Thus it misses cross beams and braces. It is drawn forward by similar energies, or more exactly, by the component of a number of attracting forces. We do not display lights, so are thus comparatively safe from discovery. They'll catch us sooner or later, though, of course." Dantus indulged in a fatalistic shrug of his shoulders as he concluded.

AT his manipulation of a number of tiny levers that were set into the control panels like the stops of an organ, the car lurched forward. Silently, swiftly, they sped on through the gloom under the great copper shell.

Through the viewing glass of a periscope arrangement that let no betraying light escape to the outside, they watched the endless lines of illuminating globes slip by beneath them. Weirdly vast and shadowy in the upper reaches, the latticed supporting columns on either side merged into continuous semi-transparent walls as the car gathered speed.

The city of Ilen-dar was left far behind. Patches of jungle flashed by; other cities. And always the endless rows of blue-white lights. There was neither night nor day in the sealed-in world; only the artificial suns that never set. Continuous subjection to the ultra-violet and visible rays of the vast lighting system was necessary to the growth and reproduction of the plant life that was so essential in keeping the atmosphere breathable.

Tommy had forgotten everything save his interest in the mechanism of the car. He and Dantus were fast friends already.

Chin in hand and eyes avoiding the pain of mourning in Tiedus' fixed gaze, Carson lost himself in gloomy meditation. As he thought back over the events of the past few days he could scarcely believe they had actually occurred or that he was sitting here in a mystery car, speeding through the rank atmosphere of an enclosed world a half billion miles from his own. Home seemed incredibly remote and desirable just then, and the future dark and forbidding.

CHAPTER V
THE TRITU ANU

BEFORE the car came to a stop Tiedus rummaged in a locker and stretched forth his hands as if carrying something delicate and fragile.

"It will be necessary for you to put this on," he said: "it will be unsafe otherwise."

Blaine stared, mystified. Was this Rulan kidding him? "Put what on?" he asked.

"A thousand pardons. I had forgotten that you do not know. I hold in my hands a cloak, an invisible thing that will hide you from the guards and from the Zara's crystal. Another secret of my father's. Dantor developed it for him only recently."

Blaine felt the texture of the stuff then; crumpled it in his fingers as its gossamer-like weight dropped in loose folds around his body. But there was nothing there: to the eyes. It simply did not exist except to the touch, and he felt no different with it on than he had before.

"Where's Tommy?" he asked suddenly, seeing that Dantus now sat alone at the controls.

Tiedus laughed. "He has been covered in the same manner," he said, "and is safely hidden from sight as are you."

Incredulous, disbelieving, Blaine called out to his friend in a tremulous voice. And Tom Farley's awed response reassured him.

"Keep close to me," Tiedus told him. The car had stopped and he directed them into the basket of the lift. The two Earth men collided violently and, clinging to each other in their ghastly invisibility, laughed crazily for a moment. As far as any observers might know there were only the two Rulans in the basket.

BLAINE fingered Pegrani's ray pistol when the cable lowered them swiftly to the roof of a huge steel cylinder that rose, a solitary and unlovely structure, in the midst of the jungle a thousand miles from Ilen-dar. The indicator informed him that seven energy charges still remained in the storage chamber of the little weapon. Its possession brought him a measure of confidence.

Careful scrutiny of the roof had shown it to be deserted, so the basket was brought to rest in a deeply shadowed portion. Immediately they stepped out, and it was sent swiftly aloft by remote control of a portable ether-wave that Dantor produced.

They encountered two of the green-bronze guards in one of the passages below and these challenged the Rulan lads with drawn pistols. The alarm was out! Fortunately Pegrani had not recognized Tiedus or all would have been lost. But the Zara was watching every Rulan community and had instructed her guards to take the Earth men into custody at all costs. Those remarkable cloaks were all that saved them. They breathed easier when the guards passed on.

Now they were in a lift, dropping speedily into the depths of the Tritu Anu. When the cage came to rest they were hustled into a maze of winding passageways that led ever downward. A wall of damp stone finally blocked their progress, but at Dantus' touch of a hidden spring a section of the solid rock swung aside to admit them to a concealed room where the lights were bright and where a delegation of Rulans awaited their coming.

With the cloaks of invisibility removed, they were welcomed by Dantor, a tall white-haired Rulan who was startlingly like his son.

They were a solemn lot, these Rulans of the older generation, but they gazed on the Earth men with sympathy and understanding. An entirely different breed from the Llotta.

This room was a secret laboratory, fully equipped for chemical and physical research. Dantor sat before a smaller replica of the Zara's crystal ball as he addressed the visitors.

"NO doubt you are puzzled," he began, using the language of the Llotta with an accent that softened its harsh gutturals, "over the calamity that has befallen you. And it is not to be wondered at. But your own danger is as nothing compared with the danger that now threatens our whole solar system. It is to explain that and to ask your cooperation in warding off the holocaust that I have sent for you.

"Since the destruction of the Tritu Nogaru we Rulans number less than one thousand, of whom three hundred are here. The Tritu Anu is foremost of the royal laboratories of Llotta-nar and its work is carried out entirely by our people. It is only on account of our superior accomplishments in science that the Llotta have allowed us to exist for so long a time, and, in this connection, I might say that the Zara has been taken severely to task for her wanton massacre of the Rulans of the Tritu Nogaru. But that is neither here nor there; it is merely a sidelight I am giving you.

"The important thing is this k-metal of yours and its relation to the plans of the Llotta. Antrid, as you know, is a dying world; coming rapidly to the end of its resources. And, as our ancestors did before us, the Llotta have been casting their eyes about for a new home. The inner planets beckoned, especially your Earth, but it was manifestly impossible to reach them as there is insufficient fuel in all Antrid to provide for the voyage of even one space ship. Then, with the long range searching rays of the crystal ball television and sound reproducers, they discovered the use of this k-metal. The sending of Antazzo to your Earth followed.

"The rest you know insofar as his activities are concerned, but what you do not know is this: The Llotta have constructed a huge steel tube that is set deep into the crust of Antrid; an enormous rocket-tube if you please, like one of those on your space ship. They plan to use the energy of this supply of k-metal in setting up tremendous streams of electronic discharges from the great tube and thus to swing the satellite from its natural orbit. They would send this entire world hurtling through space toward the inner planets, and, by proper control of the rocket discharges, bring it close to your Earth where it would become a secondary satellite at close range. Then they could war on you at their leisure and finally take Earth as their new home. Thus have the Llotta planned."

"What!" Blaine exclaimed. "Why, we'd blast them from the skies before they were started. They haven't a chance."

DANTOR nodded gravely. "I am sure of it," he agreed: "I have seen your great guns in the crystal. But they are blind to that possibility. And there are other serious flaws in the plan. The incentive, of course, lies in the certain knowledge that we are using up the internal heat of Antrid so rapidly that less than a century of life now remains to its peoples. Our power is produced by admitting water to the interior through myriads of tubes that serve the double purpose of introducing the water and conveying the generated steam back to the surface, where it produces electricity by driving great turbine generators. This electricity is distributed by charging the copper shell and the ground beneath at high frequency; it is collected from the air between by the heaters and various machines that use it. But the shortage is ever more serious and Antrid is cooling off. Thus the need for the k-metal and thus the sending of Antazzo. And now for the flaws:

"The Zara, in killing Antazzo, frustrated her own plans, as he alone, of all her people, knew how to use this marvelous energy producer. Realizing this, she set about to make friends with you two in the hope that the information might be obtained from you. That was a great mistake and raised an unexpected obstacle."

"Well, I'll be damned!" Blaine exploded. "No wonder she tried her wiles on me. Tried to make a sucker out of me, didn't she?"

Dantor smiled knowingly. "More about Clyone later," he said. "Actually she is enamoured of you, Carson, and besides she is not really responsible for the mad plan herself. But that tale can wait.

"The basic and most serious flaw in the plan is this: It can not possibly succeed, no matter how successful their attempts. What they do not understand and will not believe when I tell them is that the only result of the mad experiment will be the complete destruction of the solar system, Antrid and themselves included. Complete and horrible annihilation, I say!" Dantor paused and eyed his visitors solemnly.

IN his mind's eye, Blaine could not visualize such a thing nor picture the possible explanation. But he saw that Tommy had paled and was clenching his fists. Tommy was more of a scientist; it must be he realized what this enterprise involved.

Dantor was speaking again, in low, intense tones: "What they are refusing to see is that the delicate balance of the solar system will be disturbed if a body as large as Antrid is moved a half billion miles sunward. All bodies are kept in their orbits by a nice balance of mass attraction and centrifugal force; if a single one is altered all others are affected. What would happen is easy to calculate. First off, when Antrid approached the inner planets all bodies in the system would change their paths and the altered forces would cause severe earthquakes, tidal waves and other natural disturbances of disastrous extent.

"These would increase in violence as Antrid drew nearer to the sun, and, if she finally took up her position as a new satellite of the Earth, the entire solar system would be in chaos. By this time, even if life still remained on Earth, it would quickly become extinct, for the vastly increased tidal forces on that body would flood the land to the peaks of the highest mountains. Earth would draw in closer to the sun due to loss of velocity and increased mass of the Earth-moon system. Tremendous new forces would rend asunder the Earth, its moon, and Antrid. Venus and Mars, following suit as the forces equalized, we would have a dead universe."

TOMMY believed him. That was apparent from his furrowed brow and grim set jaw. "I'll never give 'em the secret of the k-metal," he grated. "Nor will Carson; I'll gamble on that. We'll die here before they'll get it out of us."

Blaine seconded his remarks fervently. Then, turning to the Rulan scientist, "Perhaps," he suggested, "we might remain in hiding here for an indefinite period. Perhaps even we might contrive a way of getting to the store of k-metal and regaining possession of it. They'd be licked for sure then."

Dantor beamed. "That is exactly why I sent for you," he said. Then sobering anew, he added, "But I fear that would not be the end. They will not give up. Another emissary would be transmitted to duplicate Antazzo's exploit on Earth and in five of your years the danger would again be faced. They would take infinite precaution to prevent a second failure. We must make it forever impossible—now."

"How can we? My God, it's hopeless!" Blaine groaned.

"Nothing is hopeless, my boy. Consider the plight of the Rulans. No, there is still hope and we will leave you to think it over—if you are willing. It is necessary that we Rulans show our faces above before we arouse the suspicions of the guards."

"Of course we're willing. We'll stay as long as you say—and help." Blaine was intensely earnest and Tommy chimed in with his old time fervor and enthusiasm. But hope of success seemed remote.

A murmur of approval came from the assembled Rulans, and Dantor wiped a trace of moisture from his tired old eyes. "Thank you," he said simply. "This chamber is insulated from the searching rays of the crystal spheres. You are safe for the present and will be supplied with everything you need. And I shall return shortly to discuss the matter in further detail."

THE two Earth men were alone then, in the uncanny silence of the underground retreat, regarding each other with awed comprehension. What patient, hopeless creatures these Rulans were! Knowing they were doomed, and without thought of their own safety, they were bending their every effort to the impossible task of saving the universe from the madness of the Llotta.

"What do you know about that?" Tommy said, after a while.

"It's true, what he said?" Blaine asked. "What would happen to our world, I mean—and to the rest?"

"Not a question of doubt. He's doped it out to a T. Smart guy, this Dantor."

"What do you think? Is there a chance? Think—"

"Hush!" Tommy interrupted him. "Didn't you hear something?"

The silence was ghastly; depressing. Blaine heard distinctly the beating of his own heart.

Then it was there again, that sound—a muffled scream from the other side of the stone door. A woman's scream of desperate entreaty. A shuddering, long-drawn moan, trailing off into deathly silence.

CHAPTER VI
ULANA

BLAINE was tugging at the lever he had seen the Rulans use in opening the stone door from the inside. Tommy, less excited, tried to press one of the invisible cloaks into his free hand.

"Here," he begged. "Don't be a damn fool! They'll get you, the devils."

But the great block of stone was swinging already and the young pilot squeezed through and into the passage. He stumbled over the crumpled figure of a young girl and into the arms of one of the green-bronze guards.

Recovering instantly, he prodded the big fellow's ribs with the ray pistol. "Stick 'em up!" he snarled. Then, realizing the words were meaningless to the other, he said, "Raise your hands—above your head! That's right. Stand still now, or I'll use the ray."

The guard, his face ghastly in the dim light, obeyed. But his wary eyes never left Blaine's for an instant.

A short way down the hall was the body of a young Rulan. Blaine shuddered as he saw it was headless. The ray had nearly missed that time, its energy spent before complete disintegration was effected. The girl lay still at his feet. With quick fingers he frisked the guard, finding his ray pistol and one gas grenade. What was he to do with the big fellow? He ought to let him have it, but somehow he couldn't.

Tommy was in the passageway then, invisible. The big guard stifled an amazed cry as his husky voice came out of the nothingness. These devils of Earth men! They had worked their evil magic on the Zara: had she not ordered that their lives be spared? And now there was this! His thoughts were written large on the ordinarily expressionless countenance, and Blaine was tempted to laugh at his affrighted dismay.

"Come on, you bonehead," Tommy was saying in English. "Bring the big bum inside. I'll carry the girl. Hurry; there'll be a million of them in a minute."

The girl's huddled figure was raised by unseen hands. Poised in mid-air for a moment, it floated joggily, unsteadily through the crack of the partly opened stone door. The guard, wide-mouthed and staring, muttered supplication to the war gods of Antrid.

SAFELY inside the secret chamber, the Earth men made haste to truss up the guard and gag him. He was as tractable as a child under the invisible fingers of Tom Farley, with eyes imploring the evil spirits for mercy. And when Tommy's head appeared, drifting, unsupported by a body; to be followed by arms and shoulders that seemed to materialize from nothingness, the big fellow struggled panic-stricken in his bonds, shaking with superstitious terror.

Blaine straightened the girl's limbs where she lay on a low couch. She was breathing in low shuddering gasps, but a swift examination assured him she had not been harmed. Her beautifully chiseled ivory features were fixed in an expression of nameless dread. A mass of red-gold hair tumbled in confusion about her face and shoulders and when the pilot smoothed this back his heart did a most peculiar flip-flop. Sort of jumped into his throat and stuck there. This Rulan maiden was a vision of feminine loveliness if there ever was one; a dream.

Tommy watched him with a cynical smile, and said with mock contempt, "So you're the guy who swore you'd never tangle up with a femme! Just a month ago, too. Now look: first you get this Zara woman all het up over you, and now this one's got you all het up over her. You make me sick!"

There was no fitting retort. Besides, this thing that had come to him was too serious; too big. He couldn't kid about it—even with Tom. Why, he'd always pictured this very girl in his thoughts; had always dreamed of meeting her some day. And here she was: a living, breathing reality. She was stirring, too, now; breathing easier. Her eyes opened wide; frightened, innocent ones like a child's, blue-gray and fringed with long lashes that raised dewy from the smooth ivory of her cheeks.

"ANTIUS, my brother," she exclaimed, remembering, "where is he? I saw—I thought—and the guard; he wanted to take me—oh!"

Hands fluttering to cover her face, she was sobbing now, and Blaine raised her in his arms, clumsily attempting to comfort her.

"Your brother," he said gently; "I'm afraid the guard did away with him. He is no more."

"Y-yes. I remember now; I saw." She shuddered and became still, her tousled golden head somehow finding a comfortable hollow beneath Blaine's shoulder.

And then, bravely, she sat erect and faced him. "I—I suppose I shouldn't feel so badly," she said. "We always expect it. But I was so fond of him, and he was the last. I am alone now."

"Not alone," said Blaine; "you have me—us, that is. We are the Earth men, you know. And you are safe here."

"You are Carson?" she inquired.

"Yes, and my friend is Farley. That is how your people address us, but we had rather you call us Blaine and Tommy."

Tom Farley was grinning like an idiot. Didn't he have any more sense? Blaine thought. The girl would think he was making fun of her.

"I am Ulana," she said simply.

The stone door opened silently and Tiedus slipped in, closing it swiftly behind him. He stared at the girl and at the trussed-up figure of the guard.

"So!" he exclaimed; "this is the explanation." He breathed heavily as if he had run a long way, and his face was flushed with excitement.

"Why? What's wrong?" Blaine sensed a calamity.

"The Zara—she must have seen you in the crystal. She is in a murderous rage and has visited her wrath on the Tritu Anu. Even now Dantor is on his way to Ilen-dar in answer to her summons."

"Tiedus! I'm sorry. It is my fault entirely, but—but we heard Ulana cry out."

"You did quite right, Carson. I should have done the same myself. And actually it makes little difference as far as we Rulans are concerned. We had not long to remain in this life, anyway. It is only that your hiding place might be revealed; that our plans to outwit the Llotta will fail."

"You—you think she will make away with Dantor?"

"No; he is too valuable as a scientist. But the guards are awaiting her orders to repeat what happened in the Tritu Nogaru. She depends on the work of this laboratory a great deal, though it may be she will stay her hand."

HE was fussing with the controls of the small crystal as he spoke, and it sprang into life with the peculiar shifting milkiness. Then, clearly, they were looking into the council chamber at Ilen-dar. Clyone was there, pacing the floor. Dantor had just arrived with two of the green-bronze guards. The Zara, though nervous, was curiously calm and polite in her greeting of the aged scientist.

"Dantor," she said, "I want these Earth men."

"I can not produce them, Your Majesty."

"You will not, you mean." Clyone dropped her voice. "For two reasons, Dantor, I must have these Earth men. And they must not be harmed. We need them on account of this k-metal that was brought by Antazzo, whose ugly body I so foolishly destroyed."

"Two reasons, you said, oh Clyone?" Dantor smiled knowingly.

"Yes, two!" said the Zara defiantly. "I love this Carson, if you must know. And it is the only influence for good that ever has come into my life, Dantor. Oh, can't you see? I *must* have them."

Blaine felt the hot blood mount to his temples. Tommy giggled like a moron. And Ulana drew away, ever so slightly, it was true, but still it was a definite withdrawal. Damn this leopard woman, anyway!

"He is not for you, oh Clyone," Dantor was saying, "To people of his world the very thought of such a woman as yourself is repulsive. A murderess he would call you! Their reactions to the taking of human life are entirely different from those of the Llotta. They are—you will pardon my saying it—more like those of the Rulans. The Llotta hold life cheap; they hold it dear. To your people you are not a bad woman; only a foolish one who sometimes, in the heat of passion, upsets their plans by the sudden snuffing out of a life that is valuable to those plans. Do you not see my point? He is different; to him you are the wickedest woman whom he has ever encountered—a monster."

THIS was strong talk. Blaine drew a quick breath, anticipating another of her black rages and sudden death for poor old Dantor.

But Clyone suddenly was on her knees before the old scientist, pleading with him! Creature of strange caprices! Though humanlike in her emotions when in her softer moods, she was more like the feline to which Blaine had likened her, when those soft moods had passed.

Somewhere overhead, in the chambers of the Tritu Anu, there was the sound of a muffled explosion. Its shock was felt even here in the rock-hewn secret apartment. Tiedus went white. Quickly he manipulated the controls of the crystal sphere.

"It can't be," he exclaimed. "The guards would not disobey her, and she has ordered no action."

Swiftly, then, the searching ray of the apparatus swung back to the Tritu Anu itself, boring into the vast structure above them. One of the chemical laboratories was completely wrecked; maimed and dying Rulans were everywhere in the ruins. And those who staggered to their feet were shot down by the green-bronze guards who stood at the doorway.

Then, floor after floor was revealed in the all-seeing crystal. Everywhere it was the same. Merciless, cold-blooded destruction of the Rulan scientists, the most valuable of all in the Llott scheme of things. The Earth men were speechless with horror. Ulana once more buried her head in Carson's shoulder, moaning helplessly.

The scene shifted again to the council chamber of the palace in Ilen-dar. The Zara had not risen from her knees; she was still pleading with Dantor. She knew nothing of the massacre.

"Ianito!" Tiedus gasped. "It must be he."

AND once more the view was changed. They were in the huge dome through which they had entered this mad world. Near the base of the great telescope a bullet-headed Llott was gazing into the depths of one of the crystal spheres, watching the carnage in the Tritu Anu and shouting his orders to the guards. "Slay, slay, slay!" he yelled. "Not a Rulan shall remain in all Antrid. It is Ianito commanding you, Ianito the Great, master of our destinies, Dictator Supreme. Let not one escape; I command it. Then will come the great day of release; of conquest. A new home, a new world awaits you for the taking."

"It is as I thought," Tiedus groaned: "it is the end. He has taken things in his own hands at last."

The sphere went blank at his touch of a lever. His shoulders drooped and he spread his hands in a gesture of resignation.

"What in the devil!" Tommy exploded. "Can this guy overrule the Zara? Is he that powerful?"

"He is actual ruler of the world that is Antrid; the power behind the throne. Clyone must do his bidding. He has seen that she is softening and resolved to speed things up himself."

A sudden bedlam could be heard in the corridor outside the stone door. This Ianito had gone the Zara one better. He had located them; probably saw the capture of the guard and the rescue of Ulana on the very spot where his minions now hammered for entrance.

"They will take you!" Tiedus whispered. "There is no doubt as to the orders issued by Ianito. They will take you alive and bring you to him. You will be compelled to yield the secret of the metal that energizes."

"Not on your life! We'll refuse." Blaine was very positive.

Tiedus smiled sadly. "There is the pink gas, you know," he reminded him. "No, Carson, there is but one way. You must go out into the jungle and hide for a time. Dantor will return later; it is certain he will be spared. And he will contrive some way of outwitting them. Come; there is a passage."

Blaine saw the wisdom of the argument. It was their only chance. There was a blast that shook the ground beneath their feet, and a huge section of the stone door was blown into the room. He drew Ulana close with a possessive encircling arm.

THEY were in a dark narrow passage now, following the whispered voice of Tiedus. It was damp and rankly odorous there in the darkness, and slimy things wriggled over the floor, brushing their ankles clammily. Behind them there was the roar of another explosion and the shouting of angry voices. The guards were in the secret chamber and hot on their trail.

Tiedus was fumbling with something ahead of them where they had halted; something that rattled and clanked and finally came free. A door opened into the deep-shadowed green of the jungle.

"Go now, quickly," he warned them. "Hide yourselves as far in as you dare go. You will be lost, but will later be directed by a mental message from Dantor. I shall advise him from the spirit world. We do that, you know, we Rulans. But I must leave you now; I must hold back the guards to give you time. Go, friends; farewell."

In his hand Tiedus held the ray pistol they had taken from the captured guard. He would account for a few of the Llotta at least. Blaine reached for him to restrain him; it was unthinkable that this fine lad should sacrifice himself for them. But Tiedus was gone; he had slipped away into the black depths of the passageway.

"Come on, snap into it!" Tommy grated, his voice brittle with suppressed emotion. "The kid's right; let's go." He pushed his way into the matted growth of the jungle.

Holding Ulana close and not daring to look into her eyes lest he should see what he knew was there, Blaine followed his friend. The mysterious depths of the pale green forest closed in about them.

CHAPTER VII
IN THE JUNGLE

THEY had progressed not more than twenty paces into the dense undergrowth when the gleaming wall of the Tritu Anu was entirely hidden from view. The artificial sunlight seeped through the mass of vegetation overhead, a ghostly green twilight that made death masks of their faces. But of the lights themselves, of the great latticed columns, of the enormous sponge-like blossoms of the upper surface of the jungle sea, nothing could be seen. They were deep in a tangled maze of translucent flora that was like nothing so much as a forest of giant seaweed transplanted from its natural element. The moss-like carpet beneath their feet was slushy wet and condensed moisture rained steadily from the matted fronds and tendrils above. The air they breathed was hot and stifling; laden with rank odors and curling mists that assailed throat and head passages with choking effect.

Weird whisperings there were from above and all about them. It seemed almost that the uncanny, weaving green things were alive and voicing indignant protest over the intrusion of the three humans.

Ankle deep in the rain-soaked moss, their clothing drenched and steaming, they pressed ever deeper into the tangle. All sense of direction was lost.

"Guess we'd better rest now," said Blaine, seeing that Ulana was gasping from her exertions, "They'll never trail us here."

"How about this crystal thing—the searching ray?" Tommy ventured.

"It can not follow us," the girl explained, "Certain juices of the plants provide an insulator against the ray. In fact, it was an extract of these that was used in protecting the underground laboratory we just left. We are safe now and I am very tired."

So that was the reason Tiedus had been so certain they would be safe in the jungle! Blaine had wondered about that searching ray, and now Ulana's statement had stilled his doubts. Poor kid—she was all in! Her shoulders drooped and she leaned on his arm for support. His conscience troubled him for having forced the pace in the difficult footing. They need not have come so far in.

AGLINT of light through the close packed stems caught his eye; something phosphorescent it was, shining there in the green twilight. A giant mushroom! Towering seven feet from the ground, the great umbrella-like top was aglow with sulphurous light on its under side. And, beneath its ten foot spread, the mossy carpet was dry. An ideal shelter. Here Ulana might find the rest she so sorely needed, and in comparative comfort.

She curled up beside the huge stem and, half buried in warm, dry moss, immediately fell asleep. The Earth men sat gazing solemnly at each other; speechless. In the dim distance the roar of a monorail car rose faintly at first, then grew louder and louder, only to fade away once more into the whispering silence. A steady patter of jungle rain drummed on the mushroom top.

"God!" Tommy muttered, after a while. "I'd give my right eye for a cigarette."

"Me too." Blaine was hugging his knees, nodding drowsily. "A nice rare steak with mushroom sauce wouldn't go so bad, either," he drawled.

"Aw, have a heart. I'm so sick of these vitamin pills of theirs I never want to see one again."

"Yeah, but they're better than nothing. We haven't any of those even."

"Say!" Tommy jumped to his feet in sudden remembrance. "I saw a bush, back there about fifty feet, with bunches of big red berries on it. Like grapes, they looked. May be good to eat."

"Sure, they *may* be. And then again they may be poison. We can't take any chances like that. Leave 'em alone."

TOMMY growled unintelligibly and fell to walking around their shelter with nervous strides, keeping just within the dry area and glaring savagely into the steaming jungle. Blaine smiled grimly. Nerves! Tommy always was like that; always had to be on the go and doing something. His own nerves were jumpy to-day. They were in hot water this time, for sure. Had to keep on though; they were still alive, or at least half alive; and the solar system was intact as yet. If only Tom Farley would quit his infernal tramping!

"Cut it out!" Blaine snapped peevishly. "You'll have us both bughouse. Can't you sit down and take it easy?"

Tommy stopped in his tracks. "Sorry, Blaine," he said. But he remained standing, staring off into the jungle. Then, suddenly he exclaimed, "Say, I'm going for some of those grapes, or whatever they are. I'll bring a mess of them back and we can wait till Ulana wakes up. She'll know whether they're poison or not."

"Oh, go ahead. But don't get yourself lost. Yell out if you can't find us and I'll answer."

"Okay. Don't worry about me." And in three steps Tommy was swallowed up in the undergrowth.

Blaine stole a glance at the girl and something caught at his throat. God, she was beautiful! There must be some way of getting her out of this mess. Dantor, perhaps, might show the way. He ought to be sending that message soon—a mental one, Tiedus said. Poor kid, Tiedus; gone to the happy hunting grounds now, no question of that. And he intended to advise Dantor from the spirit world. As simple as that, it was. They were game, these Rulans. Fatalists, though, and resigned to the inevitable; hopeless. But a wonderful people in a rotten world.

Soon he felt his head droop and in a moment he began to doze.

When he awoke it was to the touch of Ulana's soft fingers on his arm. "We are alone?" she asked.

"Lord!" he exclaimed, rising stiffly and rubbing the sleep from his eyes. "How long have I napped? I shouldn't have."

A swift look around the small clearing disclosed the fact that Tommy was missing. He shouldn't have let him go. A sudden panic gripped him.

"Tommy! Tommy!" he called out.

THERE was not even an echo in reply. Only the whispering of the jungle overhead and all around them. His friend was gone.

"Ulana," he said, his voice trembling, "we *are* alone. Farley is lost; swallowed up in this terrible forest."

And then, suddenly, she was in his arms. Those wondrous blue eyes, swimming in tears, looked into his own. Soft red lips, upturned, met his lips; clung there.

"I am sorry, my Carson," she said softly, when he had released her: "sorry that your good friend is lost. But perhaps," more brightly—"he has but strayed away. When the mental message comes you will be reunited. He will hear it as well as you."

Blaine shook his head. In his own heart he knew he would never see Tommy again. He had wandered too close to the Tritu Anu and had been overpowered by the green-bronze guards. Their ray pistols—he shuddered at the thought.

"I have *you* now, my Carson," the girl was saying. "Only you."

In a daze of pain and happiness intermingled, he knew he was holding her close, drawing her fiercely to him. And then, raising dull eyes to stare over the precious head and into the jungle that hid his friend, he froze with horror.

A flat serpent head with wide slavering mouth and beady eyes swayed there directly behind her. Pendant, it was, on a scaly and slimy length of undulating body that coiled high above in the matted growths of the jungle. As he watched, rooted to the spot, the great head drew back and poised, vibrating, ready to strike.

IN one quick movement he flung the girl aside and whipped out the ray pistol he had taken from Pegrani. He pressed the release and a whirring sound came from the little weapon. But no crackling blue flame sprang forth to blast this creature into nothingness. Jumping aside, he was thrown to the ground by its lashing body as the great snake struck and missed.

But the pistol was useless. Short circuited by moisture, no doubt. He crouched there, calling huskily to Ulana. She must run for it; force her way into the thick undergrowth where the thing could not reach her. She lay there, helpless with terror. Then, in a flash, she was on her feet dashing to his side. God, the huge head was poised there again! Pulsating! The glittering avid eyes upon them!

Instinctively Blaine raised the pistol just as the head darted downward. The release clicked home. And, wonder of wonders, the blue flame crackled spitefully. Exploding atoms, dazzling in the green twilight. Mighty thrashings of the huge coils high up in the tangled foliage. Crashing and tearing of great stems and rope-like tendrils. But the enormous body was headless; a dead thing in the throes of its final reflexes. Only the one charge had been spoiled; the little pistol had served them well.

He drew Ulana into the thickest of the undergrowth for protection against the tremendous lashing thing that crashed into the small clearing where the giant mushroom grew. Their shelter was destroyed. He must find another; he must be forever on guard over this girl whose hand clung so confidently to his own as they wedged their way into the thicket.

"Carson! Ulana!" A familiar voice rose above the whisperings of the jungle. A voice familiar, yet unreal; supernatural; a calm, commanding one that did not sound but echoed only in the consciousness.

"Hark!" Ulana gripped his hand more tightly. "Did you hear? It is Dantor. The message Tiedus promised."

In awed silence they waited. A tiny ball of orange fire flamed suddenly in the depths of the rushes directly before them. A sign!

"Ah, you are there!" the voice broke in. "I have your mental reactions. You will follow the orange beacon to the Tritu Anu where I await your coming. Be of good cheer, my children."

WHAT magic was this? The science of the Rulans was beyond the comprehension of the Earth man. Here was telepathy in its most perfect form. Communications from the spirit plane; the orange flame—it was all so utterly fantastic that Blaine had to look earnestly at the girl to assure himself it was not a dream. She smiled confidently.

And the orange flame was moving off into the undergrowth. They must follow its beckoning, flickering light.

It was a nightmare, that journey back through the jungle to the Tritu Anu. Dantor must be in a fearful hurry, for the orange flame moved swiftly. If they stopped a moment to rest it danced there impatiently, then receded into the green shadows until they were forced to follow for fear of losing it. Ulana's light robe was torn and sodden with moisture. The perfectly rounded ivory shoulders, bare now, were scratched and bleeding from contact with thorny protuberances that covered some of the lighter reed-like stems.

But the brave girl was uncomplaining. She clung doggedly to the Earth man's hand when they were able to walk erect: followed swiftly and unquestioningly when they were compelled to crawl or wriggle through an almost impenetrable thicket.

Once she cried out in alarm and Blaine turned back to see that the wiry tendrils of a spiny, globular plant had wound themselves around her slim body and held her fast. As he grasped her hand to draw her away, others of the tendrils curled about his wrist and he too was imprisoned. They burned the flesh, those writhing things, and tugged mightily. Ulana screamed with the pain of the many that held her in their tightening grasp.

IT was alive, this thing that grew there, a huge ball with a thousand stinging tentacles. A carnivorous plant. Even as the realization flashed across his mind he saw that the spiny sphere was opening. Split vertically, the two halves fell apart to disclose the steaming interior whose walls were lined with sharp dagger-like projections a foot in length. And the wiry tendrils were drawing them in!

Almost insane with horror, Blaine released the disintegrating energy of the weapon he still carried in his free hand. Twice he pressed its release and twice the searing blue flame spurted from the glass tube that was its muzzle.

Only a few charges remained now in the marvelous weapon but once more it had served then well. The open-mouthed plant monster vanished with the clearing of the blue vapor and the ensnaring tendrils relaxed, falling from their bodies like so many loosened cords. Blaine caught the swooning girl in his arms.

Half carrying her, he struggled on after the orange flare. The base of one of the latticed supporting columns loomed vast in the eery twilight gloom, and he leaned a moment against one of its vine-wrapped members. The girl was exhausted and hung limp in his circling right arm. Still the orange beacon danced on. If only Dantor would ease up a bit. Couldn't he give them a little time?

ON and on he staggered, ploughing through the sloppy footing and the dripping clinging greens that were everywhere in his path. Slimy fronds wrapped themselves around them, impeding his progress; clinging as if they too were alive. The whispering silence closed in on them, vast and mysterious. Menacing; awful....

And then he stumbled against a metallic wall. The curved side of the Tritu Anu! His brain cleared and courage returned with a rush. The tiny orange flame danced merrily, leading him along the wall toward the door he knew was there.

Breathing easier now, his pace quickened as Dantor's guiding light slithered along the gleaming wall. Sometimes it was almost hidden from sight by the curvature of the welded plates and he was forced into a jog trot to keep it in view.

Grimly, tenderly, he clung to the delectable creature whose soft body drooped against him.

The door! The selfsame passage through which they had escaped opened before him. Grateful even for this doubtful protection, he crossed the threshold and trudged wearily along with his precious burden. Blindly trusting in the miraculous powers of Dantor, he followed the orange beacon which now seemed to smile cheerfully as it lighted his way through the winding rock-walled tunnel.

Dazed and spent, he collapsed in the arms of the aged Rulan when he reached the end of the passage.

CHAPTER VIII
LAST OF THE RULANS

BATHED and fed and attired in dry clothing provided by Dantor, the Earth man and Rulan maiden were much refreshed and heartened when, together, they finally faced the aged scientist in the laboratory of the secret apartment. He hadn't allowed them to talk as yet.

Blaine glanced at the ragged opening where the stone door had been blown away. "We are safe from intrusion here?" he asked.

Dantor shrugged expressive shoulders. "The Tritu Anu is empty of life," he said; "a sepulchre. Those of our people who were not completely disintegrated lie blackened corpses in the chambers and corridors overhead. The gas grenades, you know. The guards went to Ianito with Farley and reported you dead: lost in the jungle from which none return."

"Farley!" Blaine shouted. "He is alive?" A wild hope sprang into being, intensified to a certainty as Dantor nodded.

"Why, yes. I thought you knew. They captured him very soon after the escape, but were unable to find you and Ulana. Ianito has mechanized him; he is in a hypnotic state of complete subjection to the Dictator. A quantity of k-metal has been taken to the laboratory at the breech of the great rocket-tube, and Farley now works there with Ianito's crew, initiating them into the mysteries of the metal's uses. Things look very bad."

"Wh-a-at!" Blaine lost his elation over the knowledge that his friend was alive. Tommy was doomed, anyway. They all were doomed. "Why did you bring us back?" he asked, turning away. Blaine felt it was better to have died in the jungle than to face this certainty of lingering torture. Ianito had triumphed; the universe was fated for utter annihilation and Ulana would suffer for weeks, perhaps months, before the final swift dissolution.

UNDERSTANDING, Dantor smiled gravely. "My boy," he said, "we still live, and while we live there is hope. That is the reason I brought you back. Tiedus' message came to me as his spirit left the body and I made haste to come here as soon as the Zara released me and I knew the coast was clear."

"What hope can there be?" Appalled by the enormity of the disaster that threatened the solar system, certain of the ultimate fate that would be meted out to Tom Farley, and convinced of their own helplessness, Blaine was gloomily unenthusiastic.

"That remains to be seen, Carson. I confess it seems impossible of remedy, but the situation must be faced and studied carefully. Insignificant as we are in the vastness of the cosmos, we may yet prove to be the ones to circumvent the mad plans of the Llotta and prevent the catastrophe which is inevitable if they succeed. We must not give up while we still breathe."

The indomitable spirit of the old scientist glistened in his keen eyes, and he stepped to the controls of the crystal sphere.

"He will not give up, oh Dantor," Ulana exclaimed loyally. "He is with us to the end. Do I speak truth, my Carson?"

Her arm slipped through his and he thrilled anew at her fragrant nearness. Give up? Never! Not with Ulana to fight for. Blaine nodded wordless agreement, silenced by the expression of Dantor's face as the crystal vibrated to a musically throbbing note.

THERE in the crystal ball was pictured a vast underground workshop somewhat like the one in the great dome through which they had entered the copper-clad world. In place of the telescope there was the butt of a gigantic cannon-like tube that towered and was lost in the shadows of the vaulted chamber. Tom Farley, moving jerkily and staring with glazed unseeing eyes, was working there with a cube of the glittering k-metal. In the open breech block of the tube was a heaped-up cone of dry soil, the material they would disintegrate in producing the blast of electronic forces. Blaine groaned as his friend called for the equivalent of a milligram of radium. Though his voice was listless and his movements uncertain, Tommy knew what he was doing and was giving away the secret, powerless to resist the command Ianito had implanted in his completely subjective mind.

"Ah," Dantor breathed: "progressive annihilation of energy: a thing we never have accomplished. You excite ordinary material such as this dry soil by means of atoms exploded from this k-metal which is in turn excited by ordinary radium that can be used over and over as the primary excitant. Am I correct?"

"You are. There are precise ratios of atomic weights to be considered, of course, but it looks as if my friend is being extremely accurate in spite of his dazed condition. Man alive! There is enough material there to provide power for the entire planet Venus for a month!"

"And enough to start Antrid from her orbit," Dantor returned. "Enough to send her on her fatal journey sunward?"

"Only for the first acceleration. A vast amount of energy is needed, Carson, since the gravitational attraction of the planet you call Jupiter is enormous. Antrid will be speeded up in its orbit and the increased centrifugal force will cause it to take up a new and larger orbit where the forces will equalize. Several charges will be required in order to free her entirely from the mother body."

"THERE'S time then!" Blaine exclaimed excitedly. "What can we do to put a stop to the thing? Something to counteract this control by Ianito; to cause Tommy to err in his proportions."

"Yes, that would do it—temporarily at least," Dantor agreed, his brow wrinkled in thought; "and there are the invisible cloaks. It is a bare chance if you want to take it. I can show you the way to this underground laboratory, and, in invisibility, you might even be able to change the ratios yourself. Yes, yes, it is a very good idea." The scientist brightened in renewed hope.

"Of course I'll chance it. When do I start?"

Dantor grinned in appreciation and Ulana looked up at him starry-eyed. "I'm going with you," she stated simply.

"Not on your life! There'll be danger. I won't have it!"

"Nevertheless, I'm going. There's another cloak and besides the danger would be greater if I were alone. Where you go I go, and if you die I die with you—gladly." She twined her fingers with his and gazed at him appealingly.

"Dantor! This can't be!" He turned to the scientist for support.

The aged scientist studied the two a little while, and then said quietly, "I'm afraid it is better as she wishes, Carson. I am unable to protect her, my boy, and there is no one else who might give her shelter. We are the last of the Rulans, she and I. The very last."

"Oh-h!" Ulana moaned, pale and distraught. "All—all are gone?"

"All, my dear. In his rage the Dictator destroyed the Tritu Deanu and the Tritu Raortu when he had finished here. Those were the last settlements remaining, you know. We alone are left behind, Ulana." Dantor bowed his head and the girl sobbed silently.

"Good Heavens!" Blaine Carson was aghast at the revelation. A monstrous deed, this last one of Ianito's. He was a fit master of a world gone mad. A monster in the twisted semblance of human form.

"HE will be searching for you, oh Dantor," the girl said with sudden conviction. She had mastered her emotions and was instantly alert to every angle of the situation.

"That is true," said the old man gravely. "For myself I have nothing to fear, of course. Though insanely jealous of my accomplishments, he maintains an armed truce with me. He dares not do otherwise as the Supreme Council is aware of his shortcomings and cognizant of my superior knowledge of science. But there is danger to you two. You must make haste."

A trembling of the ground beneath them lent added emphasis to his final words. A quick glance into the crystal told them that the initial charge was at work in the huge rocket-tube. The laboratory there at its base was in confusion indescribable, the workmen running hither and yon in the effort to escape the terrific heat that radiated from the red hot breech of the tube. They jammed the exits in their anxiety to be anywhere but near this monster source of energy whose pulsating roar drowned out all other sounds in the vast chamber.

Already Antrid was accelerating in velocity. Her vitals were wrenched and twisted, groaning in protest.

"Quick now!" Blaine was adjusting one of the invisible cloaks for Ulana. He'd *have* to take her with him. And a silent prayer for her safety was on his lips.

Invisible now, and hand in hand, they followed Dantor through the deserted passageways to the lift which carried them quickly to the roof. A drumming sound came to their ears as they stood there looking up into the blackness above the blue-white lights of Antrid. Vibrating to the tremendous roar of the rocket-tube, the copper shell emitted a constantly increasing reverberation that was like a long drawn peal of thunder on Earth or Venus. It was awe-inspiring, that sonorous bombilation; deafening.

DANTOR was fumbling with the mechanisms of the remote control which Tiedus had used in returning the basket lift to the car that had brought the two Earth men from Ilen-dar. Again and again he returned to his manipulations after peering anxiously upward. But the basket did not respond to the call. They were marooned atop the empty shell of the Tritu Anu!

"Carson! Ulana! Where are you?" the aged scientist shouted above the din, his face a tragic mask, his lips compressed with anxiety and disappointment.

They grasped him to reassure him, each taking a hand. Carson, placing his lips close to the old man's ear, inquired anxiously, "What's the trouble?"

"The car does not respond. Something has happened to the motors, probably on account of the vibration. I can do nothing."

And then, piercingly through the thunderings of the copper shell, a voice broke in—Ianito's voice. "Dantor!" it shrieked. "At last I have found you. I need your help immediately. Wait there for the monorail."

Dantor gripped them tightly to enjoin silence. Ianito had located the scientist with the searching ray and was still watching and listening at his crystal. He seemed not to know that Blaine and Ulana were there.

"Very well, oh Ianito. I shall wait," Dantor shouted.

"It is good. There is important work to be done." Ianito's words trailed off into the maelstrom of sound that swirled about them.

"He's cut off," the scientist yelled. "There is but one chance now. You must come with me, depending on absolute silence and your cloaks to deceive them. It is the only way."

ULANA clung to him there in the terrifying bedlam and Blaine's fingers strayed to the comforting butt of the ray pistol. Whatever happened there were a few charges left; blasts of energy that would serve at least to postpone the end for Ulana. Or, if worse came to worst—

The sudden rush of a monorail car high overhead interrupted his thoughts. "Close to me now!" Dantor shouted; "but have a care lest one of them touch you and discover—"

A cable-hung cage dropped swiftly to the roof and they crowded in beside the scientist. Quickly it whisked them aloft to the higher plane.

In the monorail car Blaine held the girl close, and they trod softly as they dodged the guard at the porthole and stepped into the passenger compartment. Two of Ianito's technical experts were there and a crew of at least a dozen of the green-bronze giants. Unseen by any, the couple tiptoed to the farthest corner of the compartment and took seats in a recessed section.

With a quick jerk and the rising whine of motors, the suspended vehicle started back in the direction of Ilen-dar. In earnest conversation with Ianito's engineers, Dantor affected an air of nonchalance that was artfully disarming. The Llotta suspected nothing as the car continued on its way.

And then there came an ominous grinding sound from underneath the very seat occupied by the invisible fugitives. A puff of dense black smoke followed and Ulana coughed spasmodically, uncontrollably. They were coming now, two of the green-bronze ones, to investigate. There was no escape from this narrow space. And—Ulana was gone! She had slipped from his grasp in the coughing fit and he could not find her with his wildly searching hands. Another betraying cough over there. The green-bronze ones were between them. He saw one of them draw back in amazement, then clench his fingers and twist.

The ripping sound of torn material followed and the girl's head and startled face appeared: floating there, unsupported, her body and limbs as yet invisible. But they'd found her; she was lost!

CHAPTER IX
IANITO

QUICKLY stripping the protecting cloak from her body, the green-bronze one held the struggling girl gingerly but with a grip of iron. His eyes bulged from their sockets, and the other guard staggered backward with hands outstretched as if to ward off an evil spell that might be cast by this supernatural visitant.

Blaine thrust his arm through the folds of his coat, ray pistol in hand. A crazy laugh forced itself to his lips at sight of the detached member, stretched there, tensed, drifting in mid-air. The pistol prodded Ulana's captor viciously.

"Hands off of her!" the voice behind the lone arm was snarling. "Hands off, or I fire!"

The girl slipped to the floor in a heap as the amazed guard loosed his grip. And, in the same instant, the blue flame spurted. He had not intended to press the release; it was useless anyway to battle the entire outfit. But the blood lust was upon him and a savage joy in the destruction of this beast who had dared lay hands on Ulana impelled him to turn on the other. Blindly he swung, clubbing the pistol and beating in the ghastly face that wobbled there upon the spineless, superstition-bound body.

Others were coming then, hundreds of them it seemed. The pale face of Dantor appeared for an instant in the background, through the red haze that was blinding him. He only knew he was fighting desperately, viciously, and against impossible odds. The satisfying crunch of his left fist against a leering green-bronze face was followed by an excruciating pain as one of his knuckles was driven back. Hardly knowing he had pressed the release of the ray, he was mildly astonished to see that two of the guards were enveloped in the blue vapor. Scintillant tiny sunbursts within the blue. Two less of those devils! His pistol was empty and he flung it into a grinning face; he saw the blood spurt and the face change shape, crushed beyond human resemblance.

He was down then, gasping for breath against the floor plates. The weight upon him was enormous; crushing. If only they'd quit squirming so ... and pounding ... reminded him of his old football days ... some scrimmage!

Abruptly came the blankness of insensibility.

DIMLY at first, in the painful throbbings of returning consciousness, Blaine knew he was in one of the Llott workshops where machines hummed and pounded and where many operatives were busily engaged. A cool hand stroked his aching brow and he opened his eyes. Ulana! They had spared her. Alert on the instant, he was acutely aware of the babbling of voices close at hand. Ianito was there, at the base of the huge telescope, talking with Dantor, his voice raised excitedly. The monorail crew stood by, and he noted with grim satisfaction that several of them were as badly damaged as he could wish.

His gaze returned to the sweet face that bent so near. Weakly he drew the golden head to his breast; held it there a moment, thinking, hoping, planning. Then he sat up on the edge of the low couch on which he had been placed, regarding her anxiously. Evidently they had not harmed her—as yet.

Ianito had dismissed the green-bronze ones and was approaching the couch. Dantor was with him, lagging a little and pressing a finger to his lips; shaking his head gravely to warn them. They must not speak of the plans made in the Tritu Anu; must not talk.

The Dictator was regarding them now with hard eyes. But it seemed almost that something of admiration or respect, something of human-like emotion was in his cold stare.

"Hah!" he grunted, at last. "These two are in love, Dantor. It is as you explained. It is good, and fits in with my plans to a nicety. I shall spare the life of the Earth man on account of his knowledge of the inner planets; I can use him later. The girl I shall spare for a different reason, and that fits in with my plans as well."

WHAT did he mean by that last crack, the grinning devil? A sinister intent was there, behind his smooth talk. Blaine half rose from his seat in quick anger, but the girl's gentle touch on his arm restrained him. She depended on him now and he'd have to go easy until the proper time came.

"Impetuous, aren't they?" Ianito was saying, "these Earth men. A characteristic that must get them into much trouble, even in their own world."

Laughing at him, this hell-hound! Blaine gritted his teeth.

The Dictator addressed him directly. "You are a fortunate young man," he drawled sarcastically. "You have slain several of my trusted retainers and by so doing have forfeited your right to life. But Ianito is forgiving.

Mechanized, you will be of value to me when the great day comes. And it pleases me that you are so deeply attached to the Rulan maiden; it pleases me greatly."

"Why?" Blaine snapped, a great rage consuming him. Only the pressure of Ulana's fingers held him back. He would have to control his temper or he'd make a mess of things.

"Because, my dear Carson, it will so displease the Zara."

With this cryptic remark he turned on his heel and left them. A number of his technical experts awaited him at the eyepiece of the great telescope.

Dantor whispered swiftly before following him, "Keep up your courage, Carson. A way may yet be found."

The group by the telescope was an excited one. Something had occurred which must be of great moment. It came to Blaine in a flash that the reverberations of the copper shell of Antrid had ceased. The rocket-tube was silent.

"I don't know why we shouldn't be in on this," he said to the girl. "Let's go over there and see what it's all about."

ONE of the astronomers was reporting to Ianito, referring to a sheet of calculations he held in nervous fingers. "Our orbital velocity has increased greatly," he was saying, "and the new path lies at an average distance of eighty-three erds from the mother planet. According to my figures it will require six more charges to free us from her pull and another to redirect us toward our destination."

Eighty-three erds! Practically a million Earth miles. Already they had swung out to a new orbit between those of Ganymede and Callisto. And what of the effect on the other satellites? Blaine listened carefully as the astronomer continued.

"Perturbations in the movements of the other bodies in our own system are marked, and, in the case of the first satellite, have proved disastrous. It has commenced its inward journey and soon will have fallen into the gaseous envelope of the mother body. But this need occasion us no concern; it is small and there will be stabilization of the others after the second charge is fired."

Colossal conceit! What amazing ignorance or oversight of natural laws! These Llott scientists could see no farther than their snub noses, or at least no farther than the satellite system of Jupiter. And Ianito was complimenting the astronomer on his good work!

The group broke up now and the Dictator turned to the controls of his crystal sphere. Blaine and Ulana caught the view of the underground laboratory at the base of the great rocket-tube.

All was as it had been when they first saw this chamber. The breech of the huge cannon had cooled and its massive block was open. Tommy was there, fishing the radium capsule from the powdery residue in order that it might be used in exciting the next charge. A mechanical precision marked his every motion.

"It is good," Ianito grunted, flicking a lever that cut off the view. "We are progressing nicely, thanks to the generosity of the Earth beings in providing this k-metal."

His sarcastic grin was infuriating. Dantor cast a warning look in the Earth man's direction. It wouldn't do to lock horns with this self-satisfied despot; at any rate, not now. Blaine's mien was expressionless as he faced him.

THE view in the crystal was another familiar one when Ianito made a quick readjustment: the throne room in the palace of the Zara! The Dictator snorted when he saw that Clyone was reclining lazily on her golden couch, submitting graciously to the ministrations of her handmaidens.

"Faithless creature!" he snarled. "Harlot! Parricide! But at last Ianito will have his revenge."

The hate in his voice and in those terrible glass-hard eyes was devastating in its intensity: implacable, relentless. Yet Blaine could not down the exultant feeling that came to him when he saw that this monster could suffer, too.

"What's the matter?" he sneered. "Did she throw you down?" He could have bitten off his tongue as he spoke. Ulana gasped.

And if Ianito had been in a rage before he was a madman now. Despite his contempt of the misshapen creature, Blaine quailed before the murderous glare that answered his rash words. But the Dictator was master of himself, at that; his lips tightened in a thin line and he held his peace. He actually smiled after a moment, the devil, a smile, though, of evil triumph. He turned once more to the crystal and switched on the sound mechanism.

"Clyone," he called, in velvety voice; "it is Ianito."

She looked up, startled, her chalky face gone whiter still. Her jeweled fingers fluttered to the smooth throat.

"I hear you," she replied shakily. "What do you wish of me."

"Nothing much—this time. I have visitors who request an audience with you, oh Clyone. Can you see them at once?"

"Who—who are they?" Her eyes widened at his insinuating tone.

"An Earth man—Carson—and the Rulan maiden who is to become his mate." Ianito chuckled evilly as he watched her expression.

"Carson?" she whispered, her wild eyes softening, "He—he lives?" Black hatred replaced the wondering joy that had glowed in her face. She was thinking of the statement regarding the Rulan maiden. "Why, yes," she snapped, suddenly very much alert; "I can see them. Send them immediately."

The Dictator chortled as he switched off the power. Dantor paled and looked away. So this was his scheme! He was sending Ulana to certain death at the hands of the leopard woman. Blaine bit his lips until they bled. If only he had one of their ray pistols again. If he had—

IANITO was at his side, whispering. But he couldn't see him; the devil had donned one of Dantor's invisible cloaks. Something hard pressed deep into his ribs.

"I shall be with you," the Dictator told him, "but she will not know. It is necessary, of course, that I watch over you in order that your deportment be suitable to the occasion. The death ray of Antrid is ready in my hands. Proceed, you and the Rulan maid, and see to it that you give her every attention while in the Zara's presence."

Dantor interposed an objection, "But, Ianito, you promised to spare them. I learned to love these two and want no harm to come their way."

"I keep my promise, oh Dantor. Ianito will not harm them."

"But the Zara."

"What Clyone does is none of my concern. Silence, Dantor; I command it! You will remain here." The voice of the Dictator cut like a knife.

The old Rulan scientist bowed his head and turned away. Good old Dantor! He'd done all in his power to help them. This was the end; not a question of doubt. Blaine Carson drew the Rulan maiden fiercely to him. This Clyone might meet some opposition if she attempted to wreak her spite on Ulana; she *would* meet it. There was no need for Ianito to ask that he pay every attention to the lovely, frightened girl who clung to him so trustingly.

They were in the lift then, dropping swiftly into the palace beneath the great dome that topped Antrid.

"This Clyone," Ulana whispered, "she has great power of enticement, my Carson. I fear for your loss—to me. She will take you from me, and I shall be alone—or dead. Death would be the better."

"Never!" said Blaine huskily. "Never, my dear. She has no power over me; nor will I permit her to bring suffering to you."

Ianito laughed then, an ugly cackle that came out of the unseen.

CHAPTER X
CLYONE AND ULANA

THE Zara received them in the throne room, alone. Blaine hesitated as he crossed the threshold, Ulana's trembling fingers tightly clasped in his own. The quick prod of the invisible ray pistol warned him that Ianito was at his heels.

Clyone uncurled her sinuous, black-sheathed body and rose to her feet as they neared the dais.

"Welcome, oh Carson," she purred. "Clyone has mourned you as dead, but she mourns no longer. A kind fate has returned you."

The gold-flecked eyes were all for him; it was as if she did not see his companion. Blaine fought the spell of her with all that was in him. He did not reply.

"Come to me, Carson," she pleaded, her lashes lowered. "Leave this Rulan girl and come to me."

"Where I go she goes," he replied firmly.

"Very well then," said the Zara meekly, "bring her with you. I would converse with her as with you."

Something new, this was: a gentleness Blaine had never thought the leopard woman could exhibit, even in sham. And her eyes, when she raised them, still were gentle. She extended a white arm and smiled provocatively. If this was a ruse, if she meant harm to the Rulan maid, her acting was superb. And, from what he had seen of the woman previously, he was almost convinced of her sincerity. A nature like hers was incapable of successful dissimulation. Still, he was suspicious and he shielded Ulana with his body as they came up to the throne. The Zara studied them in silence for a while. Then she spoke.

"Let me look at you, my dear," she said to the Rulan maiden.

AND Ulana, unafraid, faced her boldly. His muscles tensed, Blaine watched every movement of the Zara's straying fingers. But her gaze was direct and kindly; there was no dissembling here. It was not the same Clyone he had previously known.

"You are very beautiful, Ulana," she said softly. "Do you love this Earth man very much?"

"I do, Your Majesty."

"And you, Carson, you love her—very much?"

His answer was wordless. A sudden lump in his throat choked back the vigorous affirmative and he merely nodded, mute, as he enfolded the slight form of Ulana in his arms.

"Carson—are you sure?" Clyone was pleading, her eyes compelling; tender. Ulana drew away from his arms, waiting.

What had come over the leopard woman? She was a creature of mad vagaries, he knew, and yet this was the most convincing mood he had seen. Despite his knowledge of her past; despite his better judgment, he was drawn toward her. A step, and then quick revulsion of feeling. He recoiled and turned swiftly to Ulana.

Clyone saw and understood. Her tender mood was over in a flash and she crouched there, terrible jealous eyes fixed on the Rulan maiden. She extended a white arm with jeweled fingers, pointing. Blaine swung quickly, brushing the arm aside just as that intangible something flashed from her hand. The energy of the black disks! It had missed Ulana by inches, but crashed home—on something!

ASCREAM of terror rang out in the chamber, and there on the floor a dozen paces from the dais the thing that had been Ianito wriggled under the heap of whirring black things that suddenly covered the invisible form. He wriggled and then lay still as the angry buzzing of the black destroyers rose in triumphant, discordant song.

"Ianito!" the Zara exclaimed, thunderstruck. "He was here?"

"He was," Blaine assured her in an awed voice, "invisible, oh Zara, in a cloak contrived by Dantor, the Rulan scientist." Then blind rage overcame him. She had tried to kill Ulana; before his eyes! "You she-devil!" he roared. "I've half a mind to choke the vile life from your tainted body. Damn you! May the heat devils of Mercury burn and sear and shrivel you in everlasting torment."

She cowered as if he had struck her, and, unaccountably, he was ashamed. Cursing her like a schoolboy and using the language of the lower class Venerians!

"Please, Carson, please," she moaned; "do it. Choke me if you will and release me from my torment. I am yours to do with as you please." Throwing back her proud head, she bared her throat.

Blaine took a step forward, his knees weak beneath him.

"Carson!" It was Ulana, her hand soft on his arm.

He drew the back of his hand across his eyes. This was madness! But was ever a woman so deserving of death? Incomprehensible half-animal creature, she sat there rocking to and fro, waiting.

"No!" he said. "No! Only let us go in peace, Clyone. Your sins be on your own head. Your realization of them is punishment enough."

"Wait!" Controlling herself now, she rose once more, and her face was transfigured. Almost it seemed that she was happy. "Wait!" she repeated. "You are free to go when I have finished, but first Clyone wishes to bid you farewell."

THEY faced her in silent wonderment.

"Ianito is gone," she continued, "and the Llotta are helpless without him, unless I take over their leadership in fact. He was my master, I admit. But Clyone is able to carry on with the plans he conceived; able, but no longer willing. Clyone is abdicating. It but remains for you, Carson, to put a stop to this thing they are doing down there at the great rocket-tube. You can do it, I am certain. Go now; and think not too badly of Clyone when you have gone. Farewell."

With a quick motion she raised her fingers to her lips, then tossed a small vial crashing to the floor.

"Carson—she has taken something," Ulana stifled a hysterical sob as she spoke. "Go to her. It is the least you can do."

Blaine caught the leopard woman in his arms and lowered her gently to the luxurious cushions of the throne she had occupied for so long a time, a queen in name only. Already the gold-flecked eyes were glazing and they begged him piteously.

"Kiss me." Her lips formed the words, but no sound came.

Ulana was there, on her knees and crying. "Carson, you must," she urged him.

The spirit of Clyone, with its great burden of evil and some little of good, left the beautiful body as the Earth man pressed his lips to hers. An unwonted smile, placid and content, wreathed the still features.

The Zara was no more.

STUNNED and shaken by what they had seen, they hurried from the chamber of death. Blaine located the lift and they were quickly carried to the laboratory.

Dantor was there, working with the astronomers, and Blaine drew him aside, whispering the story in his ear in swift disjointed sentences. The aged scientist could scarcely credit his senses.

The thrumming of the copper shell to the energy of the second rocket-tube charge came but faintly to their ears in this place, since the vacuum of outer space surrounded the great domed structure. But the vibration and quakings of the satellite were transmitted to the floorplates on which they stood. They knew that Antrid was swinging ever outward from the mother planet.

"You must do it alone," Dantor was saying; "you and Ulana. I have no control over these Llotta. I am here only on sufferance of Ianito, and Ianito is no more. But they know it not. These in the dome think he is with you now, cloaked in invisibility. The tale of the cloaks has been broadcast. You are safe for the present and can descend to the base of the rocket with impunity. Ianito's name is the password. And here is a ray pistol, fully charged; two of them. He left them in his desk. Go now, quickly."

"The way—how do we get there?" Blaine's fingers closed lovingly over the butts of the pistols and he thrust them in his pockets.

"Oh yes. The lift—the one that carried you to the palace—its shaft ends deep down beneath the natural surface of Antrid in a tunnel where a moving platform will carry you to your friend. May your God and the gods of ancient Antrid be with you."

Once more they were in the cage of the lift, dropping with breath-taking speed. Down into the bowels of the satellite they sped and it seemed the shaft would never end.

THEN they were in the tunnel Dantor had told them about, smooth walls speeding past as the swiftly moving platform carried them on. The great arched chamber opened before them at last and they saw that the workmen were returning to their tasks. The huge breech of the rocket-tube had cooled to a dimly visible red, the second charge having done its work.

Hands in his pockets and walking stiffly as if mechanized, the Earth man presented himself before the guard at the entrance, Ulana pressed close to his side. He feigned the hypnotic state.

"I-an-i-to," he repeated in jerky syllables, acting the part, "he—sent us—with message—for Farley."

The guard grinned. Even here the story of the Earth man and the Rulan maiden was known. The strange leniency of Ianito in permitting them to remain together was the topic of the day. He waved them through with an indulgent gesture. Ianito knew what he was about, and would have his little joke—later.

Tom Farley was there, waiting with the Llott scientists until the breech block should have cooled sufficiently to permit them to open it and prepare the third charge. A flicker of recognition in his glazed eyes told Blaine he was not altogether gone, but Tommy gave no other outward sign. Perhaps with Ianito no longer alive, the mental control would become ineffective.

They had not long to wait, for the breech was water-jacketed and cooled rapidly. Blaine puttered around with unfamiliar test tubes and retorts, watching for a chance to get a word with Tommy in private. He was almost certain that his friend was recovering. Ulana sat there on a greasy bench, regarding the scene with anxious eyes. She was a brick: game as they made them.

Tommy was beside him then, weighing a heap of the dry soil for the next charge. "Are you all right?" Blaine whispered.

But Tom Farley stared back with not a glimmer of comprehension: He was still a victim of the mechanizing process of the Llotta. With a carefully planned but seemingly careless gesture, Blaine slid back the weight on the scale arm. This charge would be short of the proper ratio of dry soil. He wondered what the effect would be.

ONE of the Llott scientists came over then with the radium capsule, and Tommy attached it to the clamp that would hold it in contact with the cube of k-metal. The dry soil was shoveled into the breech block by the unsuspecting Llotta and the thing was ready for the placing of the excitant.

The great breech block swung home and a siren shrieked. All work in the laboratory was suspended and the workmen stood around in expectant silence. Blaine found himself worrying as to the possible result of his tampering.

"I saw you!" Tommy hissed then in his ear. "There'll be hell to pay now. We gotta beat it."

Good old Tommy! He'd recovered after all. He, too, had been shamming at the last. Blaine saw they were unobserved and thrust one of the pistols in his hand.

"Now!" his friend rasped. "Before they get wise. Grab the girl and we'll make a break for the tunnel entrance: over there."

Ulana took in the situation at a glance and was at his side. They moved swiftly in the direction of the entrance through which they had come.

A terrific roar came from the base of the rocket tube and the Llotta broke into excited screechings. Something different about it this time. There was a terrible menacing note in the jarring thump which preceded the roar. A muffled boom high in the five mile depth of rock strata above them spelled disaster of an unknown and terrifying nature. The breech of the tube was white with heat in an instant of time.

Pandemonium broke loose now and the Earth men were running for the exit to the lift, covering their retreat with brandished ray pistols. Ulana, brave girl, ran alongside, swinging a pinch bar she had picked up, ready to help.

CHAPTER XI
DISASTER

THE great crystal sphere on the central pedestal was ablaze with the scarlet warning signal of the Supreme Council. A sonorous voice from its depths boomed out above the clamor.

"Kill them! Kill the Earth men," it roared. "The Zara is dead and Ianito has vanished. Denari has mounted the throne, and it is he who commands you. Kill the traitors!"

But the Llotta and the green-bronze guards needed no command from the new ruler of Antrid. These devils from Earth had tampered with the last rocket-tube charge; probably had caused serious damage to the tube itself. They must die.

Only the guards were armed, and the Llotta swarmed so closely in pursuit of the fugitives that it was impossible for them to use their ray pistols. At the great iron gate that now closed the exit stood the guard who had admitted them. Tommy's pistol spurted blue flame and he was enveloped by the destroying energy.

Ulana screamed as a Llott grasped her, wrenching the iron bar from her hands. Blaine covered the intervening distance in a bound and his fist crashed to the fellow's jaw, snapping back his head and lifting him off his feet. He crashed to the floor plates an inert heap and the Earth man recovered the pinch bar.

Pocketing his pistol, he swung the bar with both hands in mighty circles that took terrible toll of the Llotta. They fell back before the onslaught of the infuriated Terrestrial, leaving eight of their number dead or dying with crashed skulls and broken ribs and arms.

"OPEN the gate, Tommy," he shouted. "Use your pistol on the lock if you have to." A guard was coming at him and he ducked to the floor as the blue flame crackled, singeing the hair from his head and blistering the scalp as it spent its charge in fusing a cross member from one of the steel columns nearby.

He fired from under his prostrate body and the guard thrashed his arms wildly in the blue mist, then stiffened to a sparkling vanishing figure within the dissipating vapor.

A gas grenade burst at his side and Blaine sprang to his feet, running from the spreading sulphurous cloud. The gate was open and its lock dropped molten metal. Good old Tommy!

The poison gas hid them from their pursuers for the moment and they were through the gate, all three.

"Get back!" Blaine shouted: "the gas!" He held his breath and closed his eyes as he slammed the gate and wedged it with the pinch bar he still carried. That would hold them for a while.

The gas was upon him and his skin flamed scorching hot from the contact. He mustn't breathe: mustn't open his eyes. He groped there in the scalding vapor, blindly. Tommy had him by the wrist then, dragging him away. Ulana was calling somewhere there in the darkness. His lungs were bursting. And then he knew the air was pure, and he exhaled the long pent breath noisily, and inhaled deeply.

Eyes smarting and head reeling, he saw Ulana through a haze of dancing smoke wisps he knew were illusory. She was safe, thank God! They were on the moving platform then, on the return side, and his strength was returning. Narrow escape, he'd had, from that lung-rotting gas. Ulana smiled happily when his vision cleared.

THE speeding platform carried them swiftly toward the lift that had brought them down. What if the lift would not operate? This Denari might well have shut off the power or even returned the cage to the upper end of the shaft.

"Boy, oh boy," Tommy was saying, "you sure did gum up the works. Know what happened?"

"Plenty, from the look of things," Blaine smiled grimly.

"I'll say. You cut down the dry soil ratio a third. Not sure of the exact reaction, but the expansion was too rapid. Explosion followed before the air could be driven from the tube. I'll bet the big cannon was wrecked somewhere overhead. Boy, what a blast!"

As if the last sentence were a prophecy, there came a terrific jar that twisted the platform violently from under them. They were thrown headlong and an awe-inspiring rumbling came up from the vitals of Antrid. An earthquake! The tortured satellite could not withstand the strains set up by the tremendous reactive force of the rocket-tube. The lights snuffed out and the platform came to a grinding stop. One of the underground power plants was out of commission and they were trapped here in the stifling darkness.

"Nice fix we're in now!" Tommy grunted where he had fallen.

Blaine, having located Ulana, was relieved to find that she was unharmed. "Yes," he said slowly; "but there's one thing sure: they can't follow us here unless they walk."

"Why can't we walk?" Ulana asked with forced cheerfulness. "It isn't far now."

"Oh, we can walk all right; we'll have to. And here's hoping we get somewhere." Tommy, at least, was undaunted so far.

IT was their only chance now. Blaine held fast to the girl as they felt their way along the smooth tunnel wall, and Tom Farley, behind them there in the darkness, kept up a running fire of small talk that was utterly irrelevant. Nothing could keep that Irishman down.

After what seemed like miles of steady plodding they glimpsed a light ahead. They quickened their pace. It was the open door at the base of the shaft, and the cage of the lift was there, fully lighted and waiting. Denari had not shut off the power after all. But of course! It came to Blaine in a flash; this was a private shaft, used by Ianito in his clandestine visits to the palace of the Zara and for his own use in descending to the sub-surface chamber at the base of the rocket-tube. Denari did not even know it existed.

Strange they had not been followed. Surely the Llotta could have forced that gate back there in a comparatively short time. A mass of falling rock, shaken loose by the temblor that cut off their light and stopped the moving platform, must have closed the tunnel.

They were in the cage now, shooting aloft with smooth acceleration. Tommy fidgeted and paced the floor in the narrow confines like a caged animal.

"Lord, man," he said, after a while, "what I wouldn't give for a cigarette!"

"Is that all you can think of?" Blaine was sarcastic. His own nerves were on edge. They were nearing the upper end of the shaft. "Try to do a little thinking about what's going to happen up there above Ilen-dar. We've got to do some tall figuring and some swift scrapping before we're through."

"Sure." Tommy shrugged his shoulders. "There'll be a lot of fireworks, I guess. But I wish I had a smoke just the same."

Ulana pouted. They spoke in English and she did not understand. But the expression of their faces forced a laugh to her lips, one of those silvery tinkles that caught at Blaine's heart strings. All that mattered now was to see her to safety—and happiness.

THE cage slowed up and came to rest as the automatic control of its gravity energy functioned. The door rolled back and Blaine thrust his head through the opening, pistol in hand.

There on the floor of the corridor that led to the great dome room was a crumpled figure. Dantor! It couldn't be that they had slain him! Blaine was on his knees by the body, raising the blood-smeared head with gentle hands. A deep gash extended from over the right temple up into the scalp and the skull was crushed; a mortal wound. But the doughty heart of the aged scientist still beat on, weakly, but with determination. He opened his eyes and smiled.

"Ah, you have come at last," he sighed. "I have waited here to warn you and advise you."

"Easy now." Blaine straightened the helpless limbs and cradled the drooping head on his knees. Ulana was beside him, bravely holding back the sobs that were in her throat.

"I saw—in the crystal," Dantor whispered. "And Denari struck me down when I expressed relief at your escape. Carson—Ulana—Farley—you can escape if you do as I say. Antrid is doomed; the incorrectly proportioned charge burst the rocket-tube in several places and tore the muzzle asunder where it projected from the copper shell of our world. With the explosion at the muzzle a huge section of the copper casing was blown away and the atmosphere of Antrid is now escaping rapidly into the vacuum of space...."

Dantor closed his eyes, and a spasm of pain twisted his features.

Tommy expelled a shuddering breath, solemnly expressive.

THE aged scientist fought off the grim spectre valiantly. He patted Ulana's hand as his weak voice resumed. "You will take care of her I know, Carson. Take her with you to your own world; make her happy." He fell silent once more.

"But how?" Blaine whispered.

"Oh yes, I am forgetting. The side passage—next one on the right—it leads to a storeroom of the oxygen helmets and vacuum-tight suits in which you can step forth from the adjoining airlock. Your space ship is there ... unharmed.... In it you will be able to return ... and...."

But Dantor's spirit had fled the pain-racked body. Blaine closed his lids and stretched him on the hard metal floor, crossing the thin hands on his breast. Ulana sobbed openly for a moment and then bowed her head in silence.

"The last of the Rulans," Blaine said softly, looking down at all that was mortal of the Rulan scientist.

"No," Ulana whispered, "I am the last, my Carson."

"You'll become a good American, sweetheart," he said gently. "That is, if we get away from here." There was no time to be lost, at that. At any moment this Denari might find them. "Come," he begged, drawing her from the body, "we must hurry."

Following the passage indicated by Dantor they came at last to an open door. A noticeable draft blew outward and Blaine thought grimly of the scenes that were being enacted throughout all Antrid. The air that made life possible was escaping. And the news broadcasts from Ilen-dar would have notified the entire population by this time. There would be rioting, panics, murder and suicide in the cities of the accursed Llotta and in their subject countries. A frantic effort of the scientists to stop the gap would avail them nothing: it was an impossible task now. The construction of the great shell had been a different matter; there was some natural atmosphere remaining in those days. And, finally, they would suffocate, every last one of them. They'd die miserably, purple of face and with swollen tongues protruding.

THE open door led to a railed-in balcony that looked out over the dome room. Machines still hummed there but the place was deserted save for a few scattered corpses: probably those of the Llotta who had objected when Denari usurped the throne.

A second door opened from the balcony into the store room of the moon-suits. At least these helmeted contraptions resembled the so-called moon-suits used by inhabitants of the inner planets when they visited a body having no atmosphere.

Ulana needed some assistance with the bulky equipment, and then Blaine climbed into another of the suits and locked his helmet. A moment later they were in the air-lock with Tommy, who had attired himself more quickly and was operating the controls.

At the outer hatch they waited until the air pressure reduced to a practically complete vacuum. Their suits distended ludicrously now by the pressure within, they unclamped the hatch and stepped out to the surface of the great copper shell. It vibrated under their feet to the blast from the huge gap that was not five miles distant.

The RX8 was there as Dantor had said, a slim tapered, cylinder that gleamed, a thing of beauty, in the reflected light of Jupiter which now was millions of miles distant. The sun was not visible and the light of the mother planet cast long shadows on the copper plates. Pelting ice particles clattered resoundingly against the metal helmets: frozen moisture from the escaping air of Antrid.

Blaine cried out in surprise; then remembered his companions could not hear him. There were moving shadows over there, four of them, nearing the hull of the RX8. The Llotta had beat them to it. Denari, no doubt, intending to escape with a chosen few of his subjects. He broke into a run through the now blinding hail storm. He would have to head them off; else, Ulana was lost, they all were lost.

CHAPTER XII
THE LAST OF ANTRID

TOMMY was running beside him now and Ulana was not far behind. They too had seen the danger. If they could not reach the vessel ahead of the Llotta; would not fight them off and gain possession, it was all off. They'd die here, horribly, on the roof of Antrid.

And the ray pistols were useless: they could not be fired inside the ballooning fabric of their suits without destroying it and themselves. There were only the hooks that were attached to the bulging sleeves—iron hooks for lifting—but these were heavy and sharp pointed. They might be of some use, at that.

Once they were completely blinded by a deluge of ice particles, Blaine could see neither the RX8 nor the waddling figures of the Llotta. He clung to his companions by means of the hooks, interlocking his with theirs, and waited for the storm to ease off. If ever it would! Pressing the thick glass window of his helmet against that of Ulana's, he saw that her eyes were wide with terror. But she smiled bravely and nodded encouragement. What a girl!

There was a momentary clearing a little way from the white wall and he saw the hull of the ship, a dim shape that loomed suddenly distinct and near. They dashed for the open port, still holding together.

One of the bulging, helmeted Llotta had reached the port and was scrambling inside. Blaine loosed himself and pounced on him, swinging one of his hooks in a sweeping, clawing arc. It caught in the fabric of the fellow's suit, ripping a foot-long slit. Like a punctured ballon it deflated and became a shriveled, clinging thing. The Llott hung there over the rim of the port, instantly suffocated and frozen stiff in the vacuum and intense cold of space as the air and heat of the suit was dissipated.

BLAINE dragged the rigid body from the opening and flung it to the white powdered copper surface. Wheeling, he saw that another of the Llotta had engaged Tommy. Two of them: in fact, there were three swollen figures in that mix-up. And the fourth was advancing on a smaller figure that turned and ran. Ulana! In a flash he was after them. Tom Farley would have to look out for himself, poor devil. With two of them against him, the outcome was dubious.

And then came a second snow-like deluge of white particles. He stumbled on, groping blindly; slipping, sliding in the precarious footing. It was ankle deep now, that powdery carpet of ice particles. Oh God, if that Llott devil got Ulana! He groaned aloud, a hideous mournful echo in the confines of the helmet. Groping, staggering there in the white silence, he gave up hope. The white-carpeted shell of Antrid heaved mightily from the force of some new concussion within, and threw Blaine scrambling.

Crawling now, feeling his way over the shuddering surface, he saw a dim huddled mass there in the pelting rain of ice. Moving, it was! Two bloated figures, one large and one small, rolling over and over: Ulana and the Llott who had chased her! He was there in one mad scramble and had dragged the fellow from her; was astride the rubbery inflated covering, clawing and tearing. The thing collapsed and went flat between his knees. He saw the mist of moisture-laden escaping air; felt the quick swelling and the jarring collapse as internal organs exploded from the atmospheric pressure inside the brute's body. Nauseated, he crawled away from the dead, grotesque-looking figure.

Ulana was on her knees, endeavoring to get to her feet. She had not been harmed, thanks to his good fortune in finding them. But where was the RX8? In the awful white silence, broken only by the eery patter of the ice particles on helmets and fabric, all sense of direction was lost. Through the double thickness of helmet lenses he looked into Ulana's eyes: for the last time, he thought.

AND then the white shroud lifted once more. The ship was there, not a hundred yards distant! Tommy still battled one of the Llotta, desperately circling the wary, grotesquely bobbing figure and swinging those terrible slashing hooks. The other was down, almost covered with white. Out of the picture, that one, but the remaining Llott was giving his friend a tough time of it. With the girl clinging to him, their arms hooked fast, he scuttled over the treacherous, ice-powdered copper. He had to get there quickly, and help.

Tom Farley slipped and fell heavily. The Llott was on him in a flash and they struggled madly there in splashings of white that hid them from view for a moment. Then one of them was up and the other lay still, a surprisingly shrunken and motionless figure.

The victor was coming at him then, bloated arms lashing out in swift, vicious circles. He had got Tommy, the damned swine! Blaine met his rush with a flying tackle that brought him down crashing. He lay still, the devil, knocked out probably by the metal helmet contacting with his skull. With arm poised for that slashing swing that would send him into eternity, Blaine peered through the lens of his helmet. His heart stopped beating and the upraised arm fell limp. This was no Llott: it was Tom Farley! Good Lord, he would have killed him in another second!

He tried to shake him; to bring him to. But he couldn't get hold of the bulging suit anywhere without danger of slashing it with one of those hooks. What if that fall had been fatal! Ulana was at his side now and he stared at her, white-faced, trembling in his uncertainty and horror.

And then Tommy opened his eyes. They saw him shake his head to clear it and then he, too, stared in horror. How close a call! Friend killing friend, out here in the air-less cold on the shivering shell of the dying alien world!

They helped him to his feet and through the entrance manhole. His mind awhirl with emotion, Blaine saw that Ulana was inside and then followed as in a dream. He bolted the outer cover and turned the valve that would admit air to the lock. Soon they would be inside. With their protecting coverings discarded there would be the fresh air of the interior; light; warmth. Safety for Ulana. Away from the copper-clad world, they'd be on their way—home.

ALITTLE later, Blaine Carson sat at the controls of the RX8, Ulana at his side. Tommy was below, polishing and oiling and fondling his beloved machines. The surface of Antrid was visible through the viewing port, twenty miles beneath them and receding rapidly. Swinging in its new orbit, Antrid was gasping its last.

Over there, a few miles to the east, there spouted a column of white vapor that rose from a heaped up crater of ice which extended in a circle now many miles in diameter. Heavily laden with moisture as it was, the artificial atmosphere of Antrid provided a vast storm of frozen particles as it escaped into the absolute zero of space. For many days this would continue and the pressure within would drop gradually, down, down, until the air was so rare it would no longer sustain life. And there was no hope of repairing the break: the mountain of ice prevented getting at it from outside, and the rush of air from within made the handling of patch plates and brazing torches impossible. Besides, an area of supporting columns of more than a mile diameter had been wrecked by the blast of the rocket-tube. It would require an Earth year to make such a repair, even if they could retain that atmosphere. Antrid was done for, this time.

Abruptly, Blaine turned his head from the port and gave his attention to the controls. The RX8 pointed her nose upward, away from this terrible world of disaster and death—homeward bound. With a tremendous blast from the stern rocket-tubes she headed swiftly into the heavens. A thousand miles, five, ten, they shot into space with ever increasing acceleration.

AND then a blazing orb was visible off to one side of the swiftly receding globe that was Antrid. Through the floor ports it shone, casting cheerful rays upward to the ceiling where they made a patchwork pattern of the gleaming metal.

"The sun," Ulana breathed, in awe. "I—I've never seen it, my Carson. It is most beautiful."

He drew her to him tenderly. "You'll see it every day, dear," he whispered, "when we're home."

Home—a wonderful thought! He'd not hoped to see it again; hadn't dared to since Antarro showed his hand back there in the asteroid belt. And now it was a reality. He was going home, and with him he was taking—Ulana.

"You—you think they will approve of me?" she was saying as he sent blasts from the steering rockets to swing them around on a new course sunward. "Your people, I mean. They will approve of your choice, my Carson?"

Anxiety showed in her wide-eyed gaze and she drew closer as if fearful of losing him.

If only she knew! If only he had words to tell her!

"Approve of you!" he said huskily. "Lord, girl, they'll love you! But not as I love you. It is the biggest thing—"

Tommy's discreet cough came from the head of the companionway. Blaine turned to glare savagely. His friend was standing there, grinning like an idiot and extending a paper-wrapped package.

"Look," he exclaimed guilelessly: "cigarettes. I found them, a whole carton."

"Well, I'll be damned!" Blaine exploded, careful that he spoke in English. "All you think of, all you've talked about since we left the vessel, is your hankering for a cigarette. For God's sake, get out of here and go smoke yourself to death."

But Tommy was advancing, still grinning, still extending the package. "Come on, old kid, have one," he insisted. "It'll do you good; quiet your nerves."

And his friend dropped a tantalizing eyelid. In spite of his annoyance Blaine was forced to laugh. "Oh, all right," he said, reaching for the package of smokes; "I'll take one. Just to please you. But, beat it then, will you?"

SWAGGERING as he went and casting knowing glances over his shoulder, he was gone. Great little Irishman, Tommy: always smiling, always there in a pinch, never worried, he was the best friend a man could have. They'd catch hell when they got back, for losing a part of their precious cargo. Those miserly k-metal people wouldn't give them credit for salvaging nine-tenths of the stuff (luckily only about a tenth had been removed by the Llotta): they'd only cry about the amount that was lost. And Tom Farley would laugh it off: kid them out of it.

Ulana was smiling as if she understood. She *did* understand, God bless her. She saw into this wonderful friendship and was glad. It was great to have a friend like that—and a girl like this.

Hand in hand, they gazed into the heavens before them. To the girl it was a most marvelous sight, an omen of good fortune and of happiness to come. She nestled her head into the shoulder of the Earth man as she watched; spellbound.

For a long time the silence was broken only by the steady muffled purr of the stern rocket-tubes. The aroma of cigarette smoke drifted up the companionway.

Out there in the heavens was the sun, Mars, Earth, Venus; the dear old solar system was still intact, undisturbed excepting for the slight perturbation in the region of Jupiter. Blaine doubted if the influence was measurable insofar as changes in the motions of the inner planets were concerned.

He turned to the eyepiece of the telescope and swung the instrument around to bear on the Earth. A cool green crescent was there in the field of vision: the eastern coast line of the Americas outlined clear and distinct.

"Look, dear," he whispered. "Home! Your new home is there; our home together."

She sighed happily as she gazed at the inviting sunlit outlines. "Home," she repeated, softly, reverently, "with you, oh my Carson—for all eternity."